Betrock's

ESSENTIAL GUIDE TO

PALMS

First Edition 2007

Betrock's Information Systems, Inc.
7770 Davie Road Extension
Hollywood, Florida 33024-2516

Page 1: A grouping of *Livistona* palms grace Fairchild Tropical Botanic Garden in Coral Gables, Florida.

Printed in China

Library of Congress Catalog-in-Publication Data

Leaser, David Eric.
Betrock's Essential Guide to Palms / by David Leaser.

ISBN 978-0-9629761-9-3
Library of Congress Control Number 2001012345

Contents

A World Filled with Palms

Since the beginning of time, palms have played an important role in civilization. Perhaps the earliest recorded evidence of their importance can be found in ancient Egypt where palms were cultivated for food and building materials. Edible dates were harvested for sugar, thatch was produced from fronds, and timber was milled from palm trunks. The Egyptians may have been the first culture to cultivate the common edible date palm (*Phoenix dactylifera*), whose origin remains controversial to this day. The palm was revered in Egypt and is immortalized in tomb drawings, ancient text and architecture. Massive stone columns, called palmiform columns, still stand in the Fifth Dynasty mortuary complex of Unas.

In the 18th century, Swedish botanist Carolus Linnaeus, recognizing their importance to human life, labeled them Principes, the "princes" of the plant kingdom.

For centuries, Christianity has recognized palms in rituals that are still performed today. In the fourth century, Christians began to observe Palm Sunday as the last Sunday of Lent, distributing palm fronds to church members who sometimes fashion them into small crosses.

In the 18th century, the Catholic Church introduced first non-native palms to North America. In 1749, Father Junipero Serra left Spain to help establish Catholic missions in the New World. Serra, considered the father of the California missions, introduced a variety of non-native plants to North America, including olive trees, grapevines and date palms, many of which have naturalized in California, growing in canyons and along the cliffs that border the Pacific Ocean.

While palms have been a staple of life for centuries, the search for rare and exotic palms reached a frenzy in Victorian England during the 19th century. As its empire expanded, so did England's quest for rare and exotic plants. Wealthy plant collectors commissioned "plant hunters" to search the tropics for undiscovered palm trees, leading to the identification of a myriad of new species.

In the mid-18th century, the Catholic Church established the first missions in California, bringing with them the first non-native palms to the area. Father Junipero Serra is credited with planting date palms throughout the region. Today, the Canary Island date palm (*Phoenix canariensis*) is an important part of the landscape in many regions.

At the San Luis Rey Mission in Oceanside, California, several old Canary Island date palms provide a contrast to the stark white buildings.

Above: Glasshouses, like the famous Palm House at Kew Gardens in England, allowed early plant hunters to grow exotic palms in cold climates. More than 2,500 species of palms have been discovered, ranging from the Himalayas to remote tropical rainforests.

Right: Small exotic palms, like this mottled *Pinanga disticha*, can be grown in cold winter climates if given adequate winter protection.

Back in England, gardeners housed their palms in heated greenhouses, or "conservatories." After a prohibitive tax on glass was repealed in 1845, thousands of glasshouses were soon constructed in England. The most famous of these was the massive Palm House at Kew, built by Decimus Burton between 1844 and 1848. The structure, which still stands, measured more than 24,000 square feet and could display even the tallest palm specimens.

The experiments and sheer determination of the early English plant collectors played significant roles in modern palm horticulture. Today, nearly 2,500 species have been identified, from the farthest reaches of the warm temperate to tropical worlds.

The palm kingdom boasts several unique members. The massive talipot palm (*Corypha umbraculifera*) bears the largest inflorescence (flower cluster) in the plant kingdom, the raffia palm (*Raphia* species) has the longest leaves and the dou-

Above: Built in 1913, the Botanical Building in San Diego, California was billed as the largest lath house in the world. Originally designed as a feature for the Panama-California Exposition, the "lath palace" is now framed by a variety of palms, including kentias, windmill and Bismarck palms.

Above: Early pioneers who helped forever change the world of palms include (left to right) Dr. David Fairchild, Dr. Liberty Bailey, Col. Robert Montgomery and Dr. Walter Swingle. This photo was taken in the 1930s at Montgomery's home outside Miami, Florida. The home now houses the office complex of the renowned Montgomery Botanical Center, an independent nonprofit institution devoted to advancing the science of tropical botany. Nearby, Col. Montgomery established the 83-acre Fairchild Tropical Botanic Garden, one of the finest tropical palm gardens in the world.

Right: The bright crownshaft of *Areca vestiaria* looks painted in shades of orange, yellow and green. This clumping palm is gaining popularity as a landscape plant for warm subtropical and tropical areas.

The Florida thatch palm *(Thrinax radiata)* and the silver thatch palm *(Coccothrinax barbadensis)* are native to the Caribbean, but they thrive in many tropical and subtropical climates, like Hoomaluhia Botanical Garden in Kaneohe, Hawaii. Hoomaluhia is an ideal place to see a wide variety of tropical palms and complementary plantings.

Left: Coconut palms frame the view of Diamond Head and world-famous Waikiki Beach in Hawaii. Coconut palms are not native to Hawaii; they were introduced by early Polynesians.

Right: The bright red fruit of the Norfolk Island palm *(Rhopalostylis baueri)* adds seasonal color.

ble coconut (*Lodoicea maldivica*) produces the plant world's largest fruit, weighing nearly 45 pounds. New scientific findings and fossil records estimate palms have been on earth nearly 92 million years.

Institutions like the Montgomery Botanical Center in South Florida, the Royal Botanical Gardens in England, the University of Florida and the University of California are advancing the science and horticultural knowledge of palms. The Montgomery Botanical Center maintains a seed bank program that distributes rare palm seeds to botanical gardens, universities and other institutions throughout the world.

Botanical gardens around the world are helping to introduce visitors to the vast diversity of the palm genera. From the tropics to temperate regions, public gardens are adding palm gardens and conservatories to showcase these magnificent plants.

Today, a new breed of plant hunter is introducing new palms to the market every year. Fueled by inexpensive travel and improved propagation techniques, these pioneers have sparked a mini-Renaissance in the palm world by introducing a plethora of rare and exotic species, many of which have, heretofore, been unknown in cultivation. The number of palms available in cultivation has reached an all-time high.

The coranday palm *(Trithrinax campestris)* bears stiff blue-green fronds. This slow-growing palm is able to withstand drought, wind and sub-freezing temperatures, making it a great palm for regions with harsh conditions.

The immature leaves of some species are vividly colored. These young Bismarck palms *(Bismarckia nobilis)* have a distinct purple hue. Later, these palms will bear fronds that range from green to silvery blue.

Palms make great plants for landscaping near water features. Their roots are not invasive, and unlike most trees, they don't shed small leaves and debris. This planting at the Fullerton Arboretum in Southern California includes a number of palms, including several giant fishtail palms *(Caryota maxima)*, a cliff date palm *(Phoenix rupicola)* and several *Syagrus* species.

Surprisingly, quite a few palms can survive sub-zero temperatures. The windmill, or Chusan, palm *(Trachycarpus fortunei)* can survive temperatures well below freezing. In every species, mature specimens are more tolerant of extreme weather than tender young plants. Chusan palms provide a contrast to formal gardens in England and public gardens in Vancouver, Canada.

Many palms bear colorful flowers, fronds and fruit. *Dypsis pinnatifrons* bears pink flowers, adding a strong contrast to its bright green fronds.

An archaeological dig in the ancient ruins of Masada in Israel uncovered 2,000-year-old date palm seeds. Several of the seeds were obtained by Dr. Sarah Sallon of the Borick Center in Israel, who asked botanist Elaine Solloway to attempt germination. Using a nutrient-rich medium and gibberillic acid, Solloway germinated a seedling. The seed is the oldest tree seed ever germinated.

Below: Elaine Solloway and Sarah Sallon hold the seedling, named Methuselah for the 900-year-old biblical character.

Recently, palms have been introduced that survive prolonged periods of freezing temperatures, allowing gardeners in temperate climates to enjoy their beauty in their own backyards. Palms with brilliant-colored fronds, trunks, crownshafts and flower displays are now readily available through garden centers and mail-order nurseries.

And in 2004, an exciting advance in science introduced a new, or rather "old," palm to the world. During an excavation in the ancient ruins of Masada, archaelogists uncovered date seeds that were discarded 2,000 years ago. Dr. Sarah Sallon of the Borick Center in Israel, called upon botanist Elaine Solloway to attempt to germinate a few seeds.

Solloway was able to germinate a seedling, which they named Methuselah. The seedling may be the descendant of a now-extinct species of date palm native to Israel.

The selection of palms availabe to enthusiasts has never been more extensive, providing gardeners and landscapers with endless possibilities. The palm world includes species that will only grow in the hot tropics to those that will grow in cold temperate areas.

With so many choices, many gardeners become collectors, continually seeking out the rare and unusual and testing the limits of their climate zones. Indeed, growing palms and learning more about them is somewhat addictive. The International Palm Society now boasts a membership of more than 3,000 members in 81 countries, a testament to the growing worldwide appeal of palms.

1 The largest flowers, or inflorescences, in the plant world are borne by the *Corypha* genus of palms. This talipot palm, *Corypha utan*, is topped by an inflorescence that is several feet tall. After the flowers fade, this species of palm will decline and eventually die.

2 Many palms bear crownshafts, a smooth, waxy area that connects the trunk to the fronds. The shaving brush palm *(Rhopalostylis sapida)* from New Zealand gets its common name from its erect leaves and bulging crownshaft.

3 Some palms bear spines on their trunks and fronds. This macaw palm *(Aiphanes minima)* from Cuba and the West Indies is covered with sharp black spines. The spines protect the palm from predators but also limit its use in the landscape, particularly near pathways or playgrounds.

4 The swollen trunk of the bottle palm *(Hyphorbe lagenicaulis)* from the Mascarene Islands makes this plant a great novelty in the garden. This tree is also an excellent candidate for use as a container specimen.

5 The juvenile fronds of some palms, like this flamethrower *(Chambeyronia macrocarpa)* from New Caledonia are brilliantly colored. These scarlet fronds will eventually fade to green, but they add an exciting contrast to the dense green colors of its surroundings.

The value of palm fruit

Oil from palm fruit is the most important economic product produced from palm trees. Throughout the tropical world, the fruit and seeds from the African oil palm (*Elaeis guineensis*) and coconut palm *(Cocos nucifera)* are crushed and refined and used for a variety of products, including:

- Soaps and detergents
- Margarine
- Gum
- Candles
- Plastics
- Cosmetics
- Printing inks
- Fuel oil
- Lubricants
- Textile acids

Thatch is used in many tropical countries as a roofing and siding material. Thatch is produced from the fronds of a variety of palms and makes a good waterproof covering.

Seven surprising uses

Palms play a bigger role in your day-to-day life than you may think

Around the world, entire communities rely on palm trees for their survival. From oil production plantations in Africa to coconut groves in Asia, the palm tree is one of the most useful plants on earth, second only to the grass family. Here are a few examples:

1 Food and Drink. Palms are used for a variety of food items, including sugar, dates, jelly, fruit, coconut meat and as a vegetable, called hearts of palm. The flowers of many palms are candied and sold as snacks. The fruit can be fermented and distilled as vinegar and wine or made into jelly. Coconut water is often bottled and served as a drink, called coco frio. Some food items derived from palms may require the destruction of the palm, however. The hearts of palms, called the "cabbage," are obtained by removing the growth bud, often killing the palm.

2 Shelter. An entire house can be built from a palm tree. The house can be framed with lumber milled from the trunk and the roof can be covered with thatch from palm fronds. Twine is fashioned from the leaves of many palms, including the raffia palm. Twine can also be woven into a thick, durable floor covering.

3 Medicine. Palms are used for a variety of medicines and supplements. Saw palmetto is used for prostate problems and the fruit of the betel nut palm can be used as a narcotic. During World War II, coconut water was used as a plasma substitute for soldiers who needed transfusions in the South Pacific.

4 Clothing. You can dress yourself almost completely from items derived from palm trees. Hats, sandals, purses and fabric can be woven from palm fronds. Buttons and jewelry can be fashioned from the hard seeds of several palms. Called tagua, this "palm ivory" is also carved into ornaments by villagers throughout South America.

5 Furniture. Most palm furniture is fashioned out of rattan. Rattan palms are spiny, climbing palms that grow to more than 100 ft./30 m. tall. Stems of these palms are harvested, skinned and processed to create long-lasting furniture that can be used indoors or outside.

6 Household items. Palms are used for a variety of household items, including soap, cosmetics and shampoo. Car wax can be processed from the waxy trunk and leaf coatings of several palms, including the carnauba wax palm.

7 Money. Palms are used by millions of people around the world for commerce. Entire villages in South America rely completely on palm trees for their economic survival. The most economically important palms are:

- Oil palm (*Elaeis guineensis*)
- Coconut (*Cocos nucifera*)
- Date palm (*Phoenix dactylifera*)
- Betel nut (*Areca catechu*)
- Rattan (*Calamus*)

Left: The raffia palm (*Raphia* species) produces huge fronds that are harvested to produce twine and building material. Raffia can be woven into fabric, hats, sandals and purses.

Above: Rattan furniture is created using the slender stems, or trunks, of the rattan (*Calamus*) palm. These unusual palms look more like vines than trees, climbing through rainforests with the use of hooked spines. The vine-like stems are cut down, dried and woven into beautiful, durable furniture.

Pindo jelly

The fruit of the pindo palm (*Butia capitata*) makes a great jelly. This recipe yields four eight-ounce jars.

- 2 cups prepared palm fruit juice
- 3 1/4 cups sugar
- 1/2 cup liquid fruit pectin

1. Remove pits from palm fruit, barely cover with water and bring to a boil for about five minutes.
2. Crush fruit and strain through a jellybag.
3. Measure into a saucepan. Add sugar to juice and mix well.
4. Bring to a boil over high heat, stirring constantly.
5. Stir in pectin. Bring to a full rolling boil and boil hard for one minute, stirring constantly.
6. Remove from heat, skim off foam and pour quickly into hot, clean jars.

For long-term storage, the USDA recommends heat processing all jams and jellies in a water bath canner for five to 10 minutes.

Amazing Roots

Unlike oaks and other woody trees, palms get their nourishment from a network of adventitious roots that originates from the stem rather than underground. On some palms, these roots originate some distance from the ground. These aerial roots sometimes grow into the soil, giving the palm a stilt-like appearance. Others, including the roots of many date palms, create a mass of short root stubs that extends several feet from the ground. If they reach the ground, these roots can help stabilize the palm.

The Mount Lewis king-palm (*Archontophoenix purpurea*) sports a prominent purplish crownshaft.

Palm anatomy

A brief overview of the structure of palms

When you ask most people to describe a palm tree, they envision a tall thin tree topped by a crown of green fronds. But palms come in many sizes and shapes, including shrubby bushes, climbing vines and towering trunks. Palms that dwell at the bottom, or understory, of a rainforest may mature to just a few inches tall. Others, like *Ceroxylon quindiense*, can tower more than 200 ft./60 m. in height.

Roots. All palms grow from roots, but palm roots are very different from those of other trees. Palm roots originate from the stem, or trunk, of the palm and are ad-

Flowers and fruit

Palm flowers are typically rather small, but the *Corypha* genus bears the largest flower stalk in the world, measuring up to 30 ft./10 m. tall. Most palm flowers are borne below the fronds and are followed by fruit, the largest of which are produced by the double coconut (*Lodoicea maldivica*), more than twice the size of a typical coconut. Palm flowers attract pollinators like bees and can produce significant amounts of fruit that attracts birds as well as rodents.

ventitious. Most palms produce relatively shallow root systems, making them easy to fertilize and transplant. Palms can also regenerate their root systems. Their built-in "memory" allows many palms to completely regrow severed roots. In some species, the entire root system can be removed at transplanting, making the palm easier to transport. Within months, the palm will regenerate a new root system.

Trunk. The trunks, or stems, of palms vary greatly between species. Some grow as clumping stems, others as climbing vines. Many palms bear ringed trunks, the scars from old leaves which have fallen. Others are covered in thick spines to deter predators. Some palms, like the doum palm (*Hyphaene thebaica*) will even branch, similar to a woody tree. Unlike woody trees, however, palms are monocotyledons which are not actually trees. Palms do not produce wood or bark, nor can they heal from wounds. Never attach anything to a palm that will penetrate the trunk. Nail wounds and spikes from tree trimmers create permanent and sometimes fatal damage to a palm.

Leaves. The leaves, or fronds, of palms come in many sizes and shapes. Palmate fronds are shaped like the palm of a human hand. Pinnate or feather-leaved fronds are usually deeply divided and look like an ostrich feather. Pinnate fronds can appear plumose, giving the frond a more full appearance. Several *Syagrus*, *Roystonea* and *Wodyetia* palms bear these fronds. Bipinnate fronds, seen in fishtail (*Caryota*) palms, branch off the main stem, giving the frond a unique appearance. Some palms bear entire leaves that look like saucers, although these fronds often divide if damaged by winds.

Crownshaft. Some palms bear a crownshaft, the area that attaches the leaves to the trunk. The crownshaft is a tightly packed set of leaf bases that protects the primary growth area of the palm. Most crownshafts are green, but some species bear crownshafts with striking color, including purple, yellow, red and orange.

Clockwise from top: Palmate fronds can be identified by the hand-shaped arrangement of their leaves; pinnate fronds look like a bird feather; the leaves of some pinnate fronds are arranged on various planes, giving them a plumose look.

What is not a palm?

In the nursery industry, many plants called "palms" are actually completely unrelated to true palms.

The sago palm (*Cycas revoluta*) is one of the most misunderstood. Although they bear a resemblance to palms, sago palms are actually cycads, more closely related to conifers. Ponytail palms, yuccas and aloes are also occasionally misrepresented.

This photo of a sago palm (*Cycas revoluta*) in front of a Mediterranean fan palm (*Chamaerops humilis*) illustrates the similarities between the two plants.

Portfolio of Palms

How to use this guide

How to identify and select the perfect palm

Coconuts in California?

Taken in Newport Beach, California, this photo shows the oldest northernmost coconut palm in the world. But this palm is an oddity; even though coconut palms can survive severe frosts in Sarasota, Florida, they rarely survive Southern California's sustained cool, wet climate. Microclimates, humidity, drainage and other factors play a major role in cold hardiness. A palm that survives 22°F/-7.8°C in California's dry climate may succumb to Florida's humid winters at 29°F/-2°C. This section provides "best-case scenario" temperature information. Your unique location may affect your success.

This section contains essential information on many of the most popular palms in cultivation, as a well as a few interesting rarities. Palms are alphabetized by their botanical names, such as *Phoenix dactylifera* and cross-referenced by their common names, such as date palm. Important cultural information is also listed, followed by details to help you identify or select a palm.

Hardiness. These numbers indicate the lowest survivable temperature, at least for brief periods. These number are based on the latest research, but factors such as humidity and the duration of the cold period will affect survivability. The age of the palm is also an important factor. Large, mature palms can survive cold temperatures better than young seedlings. Use this information as a starting point, but realize these numbers are not exact.

Exposure. Full sun means the palm grows best with unobstructed sunlight. Partial shade means the palm does best with at least a half a day of shade during the hottest hours. Location will affect these listings. Palms that prefer partial shade may thrive in foggy, coastal areas but require shade just a few miles inland.

Watering. Moisture needs are listed based on mature palms planted in loam. Younger palms with small root systems will need more water, as will palms planted in sandy soils.

Native to. The country of origin is listed, and many times the natural terrain (rocky slopes) is also included.

Trunk. The guide lists the mature height and spread of the palm, useful information to help you plan your garden. Trunk type is also included, whether a palm has a single trunk or multiple, clumping stems. Spines are indicated.

Leaf type. Palmate or pinnate? Use this information to help you identify a species.

General care. The portfolio also lists specific information, like preferred soil type, to ensure your palm looks its best.

Special features. Attractive features are described, like bright-colored fruit, interesting trunks or indoor potential.

Why we use botanical names

Even within small geographies, multiple names are used to describe the exact same species of palm. In Florida, gardeners and landscapers call the species *Sabal palmetto* the "cabbage palm," "palmetto palm" and "sabal palm."

When you cross geographies, the problem is compounded. Americans call *Trachycarpus fortunei* the "windmill palm," while the British call it the "Chusan palm."

This guide includes botanical names and most common names, including many found in Latin America and the Caribbean.

acoelorrhaphe wrightii

- Paurotis palm, everglades palm, silver saw palm
- Hardy to 22°F/-7.8°C
- Full sun
- Moderate water

Native to Florida and the Caribbean, this clumping palm grows to about 40 ft./12 m. tall on 4 in./10 cm. trunks and spreads to about 20 ft./6 m. wide. This palm gets its name from the silvery sheen of its palmate fronds. Common in the Florida Everglades where this palm forms large clumps of fibrous reddish brown trunks. It will form a thicket in marshy, brackish waters. It tolerates poor soil and neglect, but is prone to manganese deficiency in high pH soils. Fertilize to maintain appearance. Control spread by removing suckers which can be transplanted. Also called *hairy tom palmetto, guano preto, tasiste, tique.*

Acoelorrhaphe wrightii

acrocomia aculeata

- Gru gru palm, macaw palm, macumba
- Hardy to 25°F/-4°C
- Full sun
- Ample water

Native to Central and South America, this spiny single-trunked tree looks nearly identical to a queen palm and grows in dry, rocky regions throughout its range. Somewhat rare in cultivation, this tree bears plumose leaves and can reach 35 ft./11 m. tall with a 10 ft./3 m. spread. In its native countries, this palm is used as a food source for people and cattle. Locals use seed for palm oil and leaves for twine. Spines at eye level can be dangerous. Also called *mbocayá, mucaja, Coyol palm, corozo, tomaco, corosse, totaí.*

Acrocomia aculeata

adonidia merrillii

Adonidia merrillii

- Christmas palm, Manila palm, dwarf royal palm, adonidia
- Hardy to 26°F/-3°C
- Partial shade
- Moderate water

Native to the Philippines, this single-trunked palm grows to about 25 ft./12 m. tall with a spread of about 15 ft./5 m. A bright green crown of pinnate fronds is supported by a green crownshaft and gray trunk. This palms gets its common name from the seasonal display its red fruit provides during the winter. A very attractive tropical palm that works well as a foundation planting, this tree is a perfect choice for small yards in tropical climates. Also useful as a container specimen in a bright, indoor location. Unfortunately, this palm is highly susceptible to lethal yellowing disease, which has decimated landscape specimens in Florida. Also called *Veitchia merrillii.*

aiphanes

Aiphanes aculeata

- Hardy to 29°F/-2°C
- Partial shade
- Ample water

A large genus of 38 species of medium-sized palms native to the West Indies, Central America and South America. All have coarse, pinnate leaves and spines on their trunks and leaves and should not be planted near walkways where spines can cause injury. They are tropical in nature, although some perform well in protected subtropical locations. Some bear brightly colored flowers. Most prefer shade, but others can tolerate dry subtropical environments.

A. aculeata. MACAW PALM, RUFFLE PALM, COYURE PALM. From the rainforests of Colombia, Venezuela and Ecuador. Very tropical looking with a single trunk, which grows to 15 ft./5 m. with a spread of 5 ft./2 m. Pinnate leaves are jagged, similar to fishtail palms. Require well-drained soil. Plants bear edible orange-red to scarlet fruit. Also called *cocos rura, mararay, corozo, macagüita, marará.*

A. minima. MACAW PALM, GRIGRI, GLOUGLOU, COYORE. From the West Indies. Single-trunked palm with an arching head of pinnate fronds. To 20 ft./6 m. tall. with a spread of 5 ft./2 m. Heavily armed with spines. Prefers moist soil. Moderate to fast growth. Will tolerate low light. Good indoor potential, but be cautious of spines.

allagoptera

- Hardy to 23°F/-5°C
- Full sun
- Moderate water

This genus of small palms from South America includes two good landscape palms for subtropical to tropical climates. Trunks are short, sometimes subterranean and sometimes branching.

A. arenaria. SAND PALM, SEASHORE PALM, VESTINGA PALM. From Brazil. This trunkless species from the sand dunes of the Brazilian coast has plumose leaves that give it a bushy appearance. Leaves are green above and silvery below. The entire plant grows slowly to 5 ft./2 m. tall with a spread of 8 ft./2 m. Easy to grow in poor soils. Good for coastal areas. Also called *caxandó, côco da praia.*

A. campestris. DWARF ALLAGOPTERA, BURI, IMBURI. From Brazil, Paraguay and Argentina. Leaves are stiffer, duller than *A. arenaria.*

Allagoptera arenaria

archontophoenix

- Hardiness varies by species
- Partial shade to full sun
- Ample water

From Australia, this genus contains six highly ornamental palms. Native to tropical rainforests, all are tall single-trunked palms with prominent crownshafts and graceful, feather-leaved fronds. Leaves shed cleanly, exposing their attractive crownshafts. *Archontophoenix* are suitable for tropical and subtropical climates. These trees have naturalized on the north shore of the Big Island of Hawaii, where they grow in dense colonies. Young trees prefer shade. Difficult to transplant as crowns are fragile. All are fast growing.

A. alexandrae. ALEXANDER PALM, ALEXANDRA PALM, KING ALEXANDER PALM. Native to Australia. Hardy to 30°F/-1°C. Tall and elegant, this palm can be distinguished by its prominent ringed trunk and enlarged base. Grows to 40 ft./12 m. tall with a 15 ft./5 m. spread. Dark green leaflets are silvery gray beneath. Produces bright red fruit.

A. cunninghamiana. KING PALM, PICCABEEN PALM, BANGALOW PALM. Native to Australia. Hardy to 25° F/-4°C. This is a beautiful tropical-looking palm. Popular in Southern California where it is frequently planted. Can grow to 60 ft./18 m. tall with a 15 ft./5 m.

Archontophoenix cunninghamiana

archontophoenix

Archontophoenix alexandrae

Archontophoenix myolensis

Archontophoenix purpurea

Archontophoenix tuckeri

spread. It has a tall, single, banded trunk with a green crownshaft and pinnate leaves. Bears beautiful lilac flowers in the spring. *A. cunninghamiana 'illawarra'* is a cold-tolerant variety, able to withstand temperatures to at least 23°F/-5°C.

A. maxima. WALSH RIVER KING PALM. Hardy to 28°F/-2°C. Similar to *A. alexandrae*, but has stiffer leaves.

A. myolensis. MYOLA PALM. Hardy to 28°F/-2°C. Leaves more pendulous than others in the genus. Crownshaft is blue-green.

A. purpurea. MOUNT LEWIS KING PALM, PURPLE KING PALM. Hardy to 26°F/-3°C. Bears a purplish crownshaft, more pronounced in some specimens.

A. tuckeri. PEACH RIVER PALM. Hardy to 26°F/-3°C. Bears a lime-green crownshaft. The most tropical-looking *Archontophoenix*.

areca

Areca vestiaria

- Hardiness varies by species
- Partial shade to full sun
- Ample water

This genus contains about 60 species of tropical palms native to southern China and India through Malaysia, the Philippines, Indonesia and the Solomon Islands to New Guinea. Most grow as understory palms in rainforests. The betel nut palm (*A. catechu*) is perhaps the most famous. Locals cultivate its fruit for its narcotic properties. All are small to medium-sized, single-trunked or clumping, with feather-leaved crowns. Most require tropical conditions and a warm climate. Many adapt well to indoor and conservatory use.

A. catechu. BETEL NUT PALM, ARECA NUT PALM. Hardy to 34°F./1°C. From Southeast Asia, this single-trunked palm grows to about 40 ft./12 m. tall with a 15 ft./5 m. spread. Its tall and slender green trunk is crowned with pinnate fronds. Highly variable in appearance, betel nut palms are grown in many tropical countries for the narcotic properties in their fruit. This palm requires a warm, tropical environment with well-drained soil. Wind resistant. Makes a fine palm for the greenhouse or conservatory. Many cultivars exist, including dwarf varieties.

A. concinna. Hardy to 33°F/1°C. Native to the rainforests of Sri Lanka. This clumping palm is suited to the shady, wet tropics. To 20 ft./6

areca

m. with a 3 in./8 m. diameter bright green trunk. Often grown as a betel nut substitute.

A. ipot. HIGHLAND BETEL NUT PALM, BUNGA IPOT. Hardy to 32°F/0°C. Native to the Philippines. Similar to the betel nut palm, but shorter. This tree grows to 15 ft./5 m. tall with an 8 ft./2 m. spread. Very attractive palm that requires warm, tropical conditions and ample water. Thrives in shade. *A. macrocalyx 'Mariae'* sports a reddish crownshaft.

A. triandra. Hardy to 32°F/0°C. From Southeast Asia, this small clumping palm with slender green trunks grows to 22 ft./8 m. tall with glossy green pinnate fronds to 6 ft./2 m. long. This palm can spread to more that 10 ft./3 m. Requires filtered sun. Prefers tropical to subtropical climates. Sensitive to wind. Excellent landscape specimen. Also good as a container plant.

A. vestiaria. ORANGE COLLAR PALM, ORANGE CROWNSHAFT PALM, PINANG. Hardy to 34°F./1°C. A dramatic, attractive clumping palm from Sulawesi, Celebes and the Moluccas. Its slender crownshaft is brilliant orange, but some cultivars have red crownshafts. Others have variegated fronds. This palm grows to 20 ft./6 m. with an 18 ft./6 m. spread. Stilt roots. An excellent landscaping specimen, this palm is tropical in its requirements and requires heat, humidity and ample water. Prefers filtered sunlight. Moderate to fast growth. Also called *merah, pinang yaki.*

Areca ipot

arenga engleri

- Dwarf sugar palm, Formosa palm
- Hardy to 20°F/-7°C
- Partial shade to full sun
- Moderate water

From Taiwan and the Ryuku Islands in Japan, this clumping palm grows quickly to about 15 ft./5 m. tall and wide. This palm gets its common name from the sugar processed from its fruit. Bears fragrant, yellow flowers. Fruit produces stinging oxalate crystals which irritate the skin. This palm can be grown indefinitely in a container and can be overwintered indoors during the cold season. Pinnate leaves are blue-green with a fibrous, matted trunk. Good in full sun or as an understory palm in shady locations.

Arenga engleri

astrocaryum mexicanum

Astrocaryum mexicanum

- Chocho, lancetilla, chichon
- Hardy to 30°F/-1°C
- Partial shade
- Moderate water

From Mexico and Central America, this small, single-trunked palm grows to about 15 ft./5 m. tall with a spread of 6 ft./2 m. Young plants make good pot plants. Leaves are dark green with a silvery underside. Plants are spiny and very tropical in appearance. Good for subtropical to tropical locations.

attalea cohune

Attalea cohune

- Cohune palm, corozo, manaca
- Hardy to 30°F./-1°C.
- Full sun
- Ample water

From Mexico and Central America. A single-trunked palm to 50 ft./15 m. tall with a spread of 40 ft./12 m. Large crown of finely divided pinnate leaves. Prefers a tropical climate with adequate moisture, but will also grow in protected subtropical areas. Economic value from its seed oil and thatch. Very slow growing, but looks nice as a juvenile, shrubby plant.

bactris gasipaes

- Peach palm, pejibaye, chonta
- Hardy to 26°F./-3°C.
- Full sun to shade
- Ample water

Widely grown palm from Central and South America. Useful for its edible cabbage and fruit, which looks like a yellow peach. A clumping palm to 25 ft./7 m. tall with a spread of 10 ft./3 m. Good for temperate locations, but thrives and grows rapidly in the hot, humid tropics. Requires regular feeding. Young plants need protection from the sun. Also called *pejivalle, pupunha, chontaduro, cachipay, parépou, paripie, pisbae, paripoe, pijiguao.*

Bactris gasipaes

beccariophoenix madagascariensis

- Window palm
- Hardy to 28°F/-3°C
- Full sun
- Ample water

From Madagascar, this palm grows to 40 ft./12 m. tall with a spread of 16 ft./5 m. Single trunk with prominent rings and a crown of large, feathery leaves. Leaves often fail to separate, leaving interesting windows. Rare in cultivation, but suitable for subtropical and tropical regions. Slow growth. Excellent landscaping possibilities.

Beccariophoenix madagascariensis

bismarckia nobilis

Bismarckia nobilis

- Bismarck palm
- Hardy to 28°F/-3°C
- Full sun
- Ample water

From Madagascar. A stately palm to 30 ft./9 m. tall. Massive blue-green leaves are borne on a stout trunk. Spreads to 20 ft./6 m. wide. Tolerates sun at an early age. Ideal for subtropical to tropical locations. Many cultivars range in frond color from green to silver. Adult color is apparent in young seedlings. This tree makes an excellent focal point for large gardens. Difficult to transplant.

borassus

Borassus aethiopum

- Palmyra palm
- Hardy to 26°F/-3°C
- Full sun
- Ample water

Widely distributed genus from Africa, Madagascar, Arabia, India, Indonesia and New Guinea. These large single-trunked palms have palmate leaves and require ample space. Very slow growing.

B. aethiopum. AFRICAN PALMYRA PALM. From Africa. Grows to 70 ft./21 m. tall with a 25 ft./7 m. spread. Smooth trunk that bulges at maturity. Large crown of green palmate leaves. Similar to *B. flabellifer.* Requires a tropical location. Highly drought tolerant.

B. flabellifer. TODDY PALM, LONTAR PALM. From India. Tall palm for the dry tropics. Widespread and common in its habitat. Hard black trunk to 70 ft./21 m. tall with a 16 ft./5 m. spread. Topped with a full crown of blue-green palmate leaves. Economic value for timber, sugar and paper, which is fashioned from leaves. Large fruit, the size of a small coconut, is edible. Requires well-drained soil. Sensitive to cold. Also called *wine palm, talauriksha palm, tala palm, doub palm, Palmyra palm.*

brahea

- Hardiness varies by species
- Full sun
- Little to moderate water

Hardy fan-leaved palms native to Baja California, Mexico and Central America. Similar in appearance to *Washingtonias*. Highly adaptable for tropical to temperate locations. Very drought tolerant, but need some water in desert areas. Some will tolerate light frost. Commonly grown in the western United States.

B. aculeata. SINALOAN HESPER PALM, PALMILLA, HESPER PALM. From Mexico. Hardy to 23°F/-5°C. Grows slowly to 20 ft./6 m., spreading to 8 ft./2 m. Yellowish to gray-green fan leaves. Sparse crown. Highly drought and wind tolerant.

B. armata. MEXICAN BLUE FAN PALM, BLUE HESPER PALM. From Baja California. Hardy to 14°F/-12.2°C. Striking stiff, blue-gray, fan-shaped leaves. Slow grower to 40 ft./12 m. with an 8 ft./2 m. spread. Great landscape specimen in temperate locations. Can withstand severe frosts. Needs excellent drainage. Grow in full sun for best leaf color. Also called *Franceshi palm, blue fan palm, gray goddess.*

B. brandegeei. SAN JOSE HESPER PALM, PALMA DE TLACO, PALMA NEGRA. From Mexico. Hardy to 19°F/-7°C. in dry conditions. Dull green, dense crown of palmate fronds tinged with blue on the lower surface. To 45 ft./14 m. tall with a spread of 5-7 ft./1-2 m. Dwindling populations in nature because of harvesting for wood, cabbage and thatch. Fruit is edible when ripe and tastes similar to dates. Well suited for temperate to subtropical locations. Drought tolerant.

B. decumbens. SIERRA MADRE PALM, MEXICAN DWARF BLUE PALM. From Mexico. Hardy to 24°F/-4°C. Small clumping palm with dense green to blue-gray palmate leaves. Plant is nearly prostrate. To 6 ft./2 m. Spreads to 10 ft./3 m. Very slow growth. Excellent landscaping specimen. Needs excellent drainage. Drought tolerant. Rare.

B. dulcis. ROCK PALM, PALMA DE SOMBRERO. From Mexico and Central America. Hardy to 20°F/-6°C. Stout single-trunked palm. Slow growth to 12 ft./4 m. tall with a spread of 8 ft./2 m. Highly adaptable and well-suited for sunny, dry regions. Very attractive. Also called *suyate, capulín, soyate.*

B. edulis. GUADALUPE PALM. From Guadalupe Island near Mexico. Hardy to 20°F/-6°C. Similar to *B. armata* but with light green, fan-shaped leaves. Slow growth to 30 ft./9 m. Highly adaptable for sunny temperate to dry tropical regions. Performs poorly in damp, humid areas. Drought tolerant. Fruit is edible.

B. moorei. DWARF ROCK PALM, POWDER PALM. From Mexico. Hardy to 14°F/-10°C. Small palm to less than 6 ft./2 m. with an underground trunk. Dark green fronds have white undersides.

B. pimo. From Mexico. Hardy to 25°F./-4°C. Uncommon palm with a stout trunk to 12 ft./4 m. with a 6-8 ft./2-3 m. spread. Dull green crown of palmate leaves. Slow growing.

Brahea armata

Brahea aculeata

Brahea edulis

Brahea decumbens

burretiokentia

Burretiokentia viellardii

- Hardiness varies by species
- Partial shade
- Ample water

Attractive, tropical-looking palms from New Caledonia. All have single-trunked trunks, pinnate fronds and prominent crownshafts. Suitable for subtropical to tropical climates.

B. hapala. From New Caledonia. Hardy to 27°F./-3°C. Grows to 25 ft./8 m. with a spread of 14 ft./4 m. Has a prominent brownish green crownshaft and a crown of feathery, pinnate leaves. Nice palm for subtropical to tropical locations. Easy to grow.

B. viellardii. From New Caledonia. Hardy to 32°F./0°C. Single-trunked palm to 25 ft./8 m. tall with a spread of 14 ft./5 m. Glossy green to brown trunk with prominent rings and a sparse crown of pinnate leaves. More difficult than *B. hapala* to grow.

butia

Butia capitata

- Hardiness varies by species
- Full sun
- Little to moderate water

Native to Brazil, Paraguay, Uruguay and Argentina, this genus of eight species contains relatively stout, single-trunked palms with arching, pinnate leaves. Leaves are typically blue-green. Very popular in cultivation because of their ornamental qualities. Easy to grow, tolerating a range of soils. All require good drainage. Most are drought tolerant. Sometimes hybridized with *Jubaea* and *Syagrus*.

B. archeri. COQUEIRINHO DO CAMPO. From Brazil. Hardy to 20°F./-7°C. Small, single-trunked palm which grows in the dry grasslands in its native country. Trunk is typically subterranean, but some reach 3 ft./1 m. tall with an equal spread. Blue-green pinnate leaves.

B. capitata. PINDO PALM, JELLY PALM. Native to Brazil, Uruguay, where it grows in arid areas. Hardy to 10°F./-12.2°C. Grows to 20 ft./6 m. with a spread of 12 ft./4 m. A highly variable feather palm, with some specimens exhibiting silver-blue upright fronds, others exhibiting graceful arching fronds. The yellow to red fruits are very decorative and edible, with a flavor that tastes like mango, strawberry and banana. The edible pulp can be used for jelly or fermented to make wine. *B. capitata var. strictior* has stiffly upright fronds. Also called *cabegudo, Butiá-da-praia, wine palm.*

B. eriospatha. WOOLY BUTIA, WOOLY JELLY PALM. Native to Brazil. Hardy to 10°F/-12.2°C. Similar to *B. capitata*. Slow growth to 15 ft./5 m. with a spread of 12 ft./4 m. Single trunk with arching, pinnate leaves. Green or blue-green fronds. Highly adaptable. Drought tolerant. Rare in cultivation. Also called *Butia-da-serra*.

B. microspadix. Native to Brazil. Hardy to 20°F/-7°C. Trunkless palm with erect blue-green fronds to 3 ft./1 m. Rare in cultivation.

B. paraguayensis. YATAY PALM, JELLY PALM. Native to Paraguay, Argentina and Brazil. Hardy to 20°F./-7°C. Slow growth to 10 ft./3 m. with a spread of 10 ft./3 m. Stout trunk with graceful arching fronds. Looks like a dwarf *B. capitata*, but leaves are more green. Highly drought tolerant. Suitable for temperate to tropical climates. Also called *yatay enano, butia*.

B. yatay. YATAY PALM, JELLY PALM. Native to Brazil, Argentina and Uruguay. Hardy to 15°F/-9°C. Slow growth to 35 ft./11 m. with a spread of 10 ft./3 m. Similar to *B. capitata*, but taller and thinner. Also called *palma yatay*.

Butia capitata 'strictior'

Butia paraguayensis

calamus

Calamus caryotoides

Calamus spines

- Rattan palm
- Hardiness varies
- Shade
- Ample water

Widely distributed throughout tropical regions of the world. Most are climbing, vine-like palms with tough, spiny stems and pinnate leaves. Often grouped together as rattan palms because of their use in furniture-making. Rarely grown as landscape specimens because of their sharp spines, but some species make ideal indoor pot plants. Need a nutrient-rich soil.

C. australis. LAWYER CANE. From Australia. Hardy to 26°F./-3° C. Vigorous climber to 20 ft./6 m. with a spread of 4 ft./1 m. Vicious recurved spines. Young plants, with fewer spines, are suitable as container plants. Requires warm, wet climates in full shade. Can survive in protected temperate and subtropical climates.

C. caryotoides. FISHTAIL LAWYER'S CANE. Hardy to 18°F/-8°C. Very slender cane-line stems to 20 ft./6 m. with distinctive fishtail leaflets. Spreads to 3 ft./1 m. Usually seen as a single-trunked plant, but occasionally suckers. Thrives as an understory palm. Good pot plant. Slow growth speeds with age. Good for temperate to tropical locations. Damaged by salt air.

C. ciliaris. From Indonesia. Young plants are attractive. Common in English glasshouses. Sensitive to cold.

C. moti. YELLOW LAWYER CANE. From Australia. Fast-growing palm. Leaves are dark green and armed with yellow spines.Requires a warm temperate to tropical climate. Young plants are attractive and make good pot plants.

C. muelleri. SOUTHERN LAWYER CANE. From Australia. Hardy to 32°F/0°C. Clumping palm with spiny, climbing stems. Grows to 32 ft./11 m. with a 6 ft./2 m. spread. Less spiny than many rattans. Requires a temperate to tropical climate. Can tolerate some frost.

C. ornatus. BLACK MOUNTAIN RATTAN, ELEPHANT RATTAN. From Philippines, Malaysia, Thailand and Indonesia. Attractive climbing palm, with dark green leaves and spiny stems. Bears edible fruit. Can be trained to clump with some pruning. Can grow to 200 ft./61 m. tall. with a spread of 15 ft./5 m. Requires a tropical to subtropical location.

C. radicalis. VICIOUS HAIRY MARY. From Australia. Heavily armed palm forms dense clumps. Rarely cultivated, but requires a tropical to subtropical location.

C. reyesianus. APAS PALM. From the Philippines. Suckering palm forms spreading clumps. Long leaves are bright green with brown spines. Prefers a tropical climate. Attractive when young.

C. viminalis. From India and Malaysia. Vigorous climber to 25 ft. /8 m. Thick stems bear pale green leaves and are armed with spines. Tropical in its requirements with shade and ample water.

calyptrocalyx hollrungii

- Hollrung's palm
- Hardy to 32°F/0°C.
- Partial shade
- Ample water

From New Guinea, this small understory palm clumps with thin stems and small crownshafts. Grows to 8 ft./2 m. tall with a 6 ft./2 m. spread. Pinnate leaves are reddish to maroon when first emerging, fading to a deep green. Rarely cultivated, but has excellent potential as landscape specimens because of its ornamental traits and small size. Many would make good container palms.

Calyptrocalyx hollrungii

carpentaria acuminata

- Carpentaria palm, carpy
- Hardy to 30°F./-1°C.
- Partial shade to full sun
- Ample water

From Australia. Fast growing to 30 ft./9 m. with a spread of 10 ft./3 m. Tall, single-trunked slender palm with an attractive crown of arching pinnate fronds. Trunk is gray with prominent rings. Ideal landscaping palm as an individual specimen or in a grove. Tolerant of poor-draining, heavy clay or marl soils. Young plants are cold and sun sensitive.

Carpentaria acuminata

carpoxylon macrospermum

Carpoxylon macrospermum

- Hardy to 32°F/0°C.
- Partial shade to full sun
- Ample water

From Vanuatu, this palm grows to 90 ft./30 m. tall. Single-trunked palm with prominent white rings and a crown of large, feathery leaves. Bright green crownshaft can be 5 ft./2 m. in length. Bulging trunks of young trees are a striking green color, fading to gray as the palm matures. Rare in cultivation, but suitable for subtropical and tropical regions. Slow growth. This is a rare palm recently discovered in cultivation with excellent landscaping possibilities because of its highly ornamental qualities.

caryota

Caryota mitis

- Fishtail palms
- Hardiness varies by species
- Partial shade to full sun
- Ample water

From Southeast Asia. This genus of 12 species is easily recognized by its unusual fishtail-shaped leaflets. Fronds are bipinnate. Some grow as single-trunked palms, others as clumping specimens. Most are short-lived, dying after a few decades. All are monocarpic; individual stems die afer flowering. Fishtail palms serve a wide variety of uses in their native lands, being used for sago, fiber, sugar and palm wine. Suitable for subtropical to tropical locations. Fast growing and ideal as landscape specimens. Fruit produces stinging oxalate crystals which irritate skin.

C. maxima. GIANT TURTLE PALM, BLACK PALM. From Southeast Asia. Hardy to 27°F/-3°C. Large single-trunked palm with arching bipinnate leaves, grows quickly to 50 ft./15 m. with a 6 ft./2 m. spread. Excellent landscape palm for large lawns and parks.

C. mitis. CLUSTERING FISHTAIL PALM, BURMESE FISHTAIL PALM. From India and Southeast Asia. Hardy to 28°F/-2°C. Grows to 15 ft./5 m. with a spread of 16 ft./5 m. Clumping palm that produces a dense mass in the landscape. Light green foliage. Seed clusters are stunning and look like dreadlocks. Perhaps the most common fishtail palm in cultivation. Suitable for warm temperate areas and indoor container gardening. Needs partial shade. Responds well to fertilizer. A rare variegated form exists. Also called *tufted fishtail palm*.

C. no. GIANT FISHTAIL PALM. From Borneo. Hardy to 32°F/0°C. Tall, majestic single-trunked palm to more than 60 ft./18 m. with a spread of 18 ft./6 m. Thick gray, bulging trunk. Fronds may grow to 13 ft./4 m. long with an equal spread. One of the largest of the *Caryotas*. Excellent landscape or avenue palm, now rare in the wild as it is harvested for its cabbage. Fast growth, especially in the wet tropics. Good for tropical to subtropical regions. Makes a good indoor container specimen when young. Requires partial shade.

C. obtusa. BLACK FISHTAIL PALM. From Thailand and India. Hardy to 27°F/-3°C. Tall, single-trunked palm with dark trunk to 60 ft./18 m. with a spread of 18 ft./6 m. Similar to *C. urens*. Attractive leaves are more blunt than other *Caryotas*. Suitable for temperate regions. Requires partial shade. Good indoor specimen.

C. ochlandra. CHINESE FISHTAIL PALM. From China. Hardy to 27°F/-3°C. Single-trunked palm that grows fast to 40 ft./12 m. with a large crown of dark green, bipinnate leaves. Spreads to 16 ft./5 m. Perhaps the most cold tolerant of the *Caryotas*. Can tolerate frost. Easy to grow. Suitable for temperate to tropical regions. Prefers partial shade.

C. rumphiana. FISHTAIL PALM, GIANT FISHTAIL PALM. From Moluccas, New Guinea and Australia. Hardy to 30°F/-1°C. Single-trunked palm to 50 ft./15 m. with a spread of 16 ft./5 m. Attractive, stiff leaflets. Fast grower. Suitable for tropical to subtropical climates. Prefers partial shade.

C. urens. WINE PALM, TODDY PALM, JAGGERY PALM. From Burma, India and Malaysia. Hardy to 20°F/-7°C. Single-trunked palm with thick gray trunk to 50 ft./15 m. with a spread of 14 ft./4 m. Large crown of arching, bipinnate fronds. Fast grower that responds well to fertilizer. Tolerates full sun. Good specimen for wet tropical and subtropical locations. Good indoor plant when young. Relatively cold tolerant. Grown for its sugary sap (toddy), which is collected by cutting off the flower clusters and milking the wound. Also called *toddy fishtail palm, solitary fishtail palm, sago palm, kitul palm.*

Caryota obtusa

Caryota urens

ceroxylon

Ceroxylon alpinum

- Wax palm
- Hardiness varies by species
- Partial shade to full sun
- Ample water

From the Andes Mountains in South America, this genus contains some of the tallest palms in the world. Slow-growing single-trunked palms with pinnate leaves. Trunks are covered with a thick layer of wax which is used commercially.

C. alpinum. WAX PALM. From Venezuela, Colombia and Ecuador. Hardy to 28°F/-2°C. Tall palm to 100 ft./30 m., but more commonly to 40 ft./12 m. in cultivation with a spread of 25 ft./8 m. Leathery leaves. Prefers cool frost-free climates.

C. quindiuense. QUINDIO WAX PALM. From Colombia. Hardy to 25°F/-4°C. The national tree of Colombia. One of the tallest trees in the world; to 200 ft./61 m. in its native habitat, more commonly to 60 ft/18 m. in cultivation with a spread of 30 ft./9 m. Leaves are whitish underneath. Cultivated as a source of wax. Requires a cool subtropical climate. Also called *Andean wax palm, palma de cera, wax palm.*

C. ventricosum. From Ecuador and Colombia. Hardy to 21°F/-6°C. Slow grower to 60 ft./18 m. or more with a spread of 12 ft./4 m. Leaves are plumose. Good palm for warm temperate areas.

chamaedorea

Chamaedorea klotzschiana

- Hardiness varies by species
- Full shade to partial shade
- Ample water

Attractive feather palms from Mexico through South America. Most are small clumping palms with multiple, thin trunks; some are single-trunked palms growing to about 20 ft./6 m. tall. Excellent container specimens that can be over-wintered indoors and planted in the garden for the warmer months.

C. cataractarum. CASCADE PALM, CAT PALM, CATARACT PALM. Native to Mexico. Hardy to 25°F/-4°C. Clumping palm to 5 ft./2 m. with a spread of 8 ft./2 m. Dense and shrubby with multiple stems. Rich, glossy green leaves can be used in floral arrangements. Excellent understory palm or border plant. Good indoor container specimen. Sensitive to salt buildup. Also called *guayita de los arroyos.*

C. costaricana. BAMBOO PALM, PACAYA. From Central America. Hardy to 25°F/-4°C. Clumping palm that grows quickly to 10 ft./3 m. with a spread of 5 ft./2 m. Can be used as a hedge or screen. Similar to *C. microspadix* and *C. seifrizii.* Excellent container specimen. Also called *Costa Rican bamboo palm, pacayita, tenera.*

C. elegans. NEANTHE BELLA, PARLOR PALM, CHATE. Native to Mexico and Central America. Hardy to 30°F/-1°C. Grows as a single-trunked palm to 10 ft./3 m. tall with half-inch/1 cm. stems. More commonly seen grown as a clump with multiple plants. One of the most common houseplants in the world. Easy to grow in low light areas.

C. ernesti-augustii. TUNA-TAIL PALM. Native to Mexico and Central America. Hardy to 22°F/-6°C. Grows to 10 ft./3 m. tall with single half-inch stems/.1 cm. Looks best in a grouping. Leaves are undivided.

C. fragrans. SANGAPILLA. Hardy to 25°F/-4°C. From Bolivia and Peru. Slender, clumping palm with cane-like trunks to 8 ft./2 m. tall. Dark green v-shaped leaves. Fragrant flowers have a spicy scent.

C. glaucifolia. KIBA. Hardy to 30°F/-1°C. From Mexico, this palm grows to 15 ft./5 m. tall with single, thin stems.

C. hooperiana. Native to Mexico. Hardy to 25°F/-4°C. Clumping palm grows slowly to 12 ft./ 4 m. with a spread of 4 ft./1 m. Tolerant of low light and dry air. Makes an attractive container plant.

C. klotzschiana. Native to Mexico. Hardy to 23°F/-5°C. Single-trunked palm that grows to 12 ft./4 m. tall. Dark green trunk with white rings. Prominent crownshaft. Irregular fronds on sparse crown.

C. metallica. METALLIC PALM, MINIATURE FISHTAIL PALM, METALLICA PALM. Native to Mexico. Hardy to 25°F/-4°C. To 5 ft./2 m. tall. Undivided leaves have an unusual metallic sheen. Looks best when planted in groups. Easy to grow as a container palm.

C. microspadix. BAMBOO PALM, HARDY BAMBOO PALM, PALMILLA. Native to Mexico. Hardy to 20° F/-7°C. Excellent clumping palm to 10 ft./3 m. tall. Spreads to 6 ft./2 m. Stems have a bamboo-like appearance. Dull green leaves. Can be used as a screen.

C. plumosa. Hardy to 23°F/-5°C. Native to Mexico. Hardy to 28°F/-2°C. Single-trunked palm to 20 ft./6 m. tall with a 3 in./8 cm. trunk. Sun tolerant. Looks like a miniature *Syagrus romanzoffianum*.

C. pochutlensis. CANELILLA. Native to Mexico. Hardy to 25°F/-4°C. Slender clumping palm to 10 ft./3 m. tall with bamboo-like stems. Leaves are shiny. Can endure light frosts.

C. radicalis. COVER PALM, RADICALIS PALM. Native to Mexico. Hardy to 20°F/-7°C. Can recover from temperatures down to 9°F/-13°C. Slender, single-trunked palm to 10 ft./3 m. tall, but many are trunkless and much shorter. Relatively cold and drought hardy.

C. seifrizii. BAMBOO PALM, REED PALM, CHIATE. Hardy to 30°F/-1°C. From Central America and Mexico. Clumping palm to 10 ft./3 m. tall.

C. stolonifera. Hardy to 30°F/-1°C. Native to Mexico. Multiple-trunked palm to 3 ft./1 m. tall. Spreads by underground rhizomes.

C. woodsoniana. Native to Mexico, Central and South America. Hardy to 25°F/-4°C. Single-trunked to 20 ft./6 m. tall. Needs shade.

Chamaedorea radicalis

Chamaedorea microspadix

Chamaedorea elegans, known as 'Neanthe Bella' in the horticultural trade, is one of the most commonly grown houseplants in the world.

chamaerops humilis

Chamaerops humilis

- Mediterranean fan palm, European fan palm
- Hardy to 0°F/-17.8°C.
- Full sun
- Moderate water

Native to Southern Europe and Africa where it grows in coastal areas, often on rocky slopes. Can sustain temperatures of 0°F/-18°C for short periods. Extremely cold tolerant, able to withstand heavy frosts. Tolerates wind, drought and poor soil, but requires good drainage. Highly variable species grows slowly to 20 ft./6 m. tall and sometimes as wide. Bears palmate fronds on spiny stalks. Several cultivars are grown, including single-trunked and clumping forms in various shades of green and bluish green. Trunks are covered with a hairy fiber. A variegated cultivar is also grown. Very widespread in cultivation. Ideal plant as a focal point or as an impenetrable hedge. Ideal as a container plant, especially when young. Cultivated varieties include:

'green mound.' Bright green; forms a dense clumping mound.
'elatior.' Bears a single, tall trunk.
'cerifa.' Beautiful blue-gray variety from Morocco.

chambeyronia

Chambeyronia macrocarpa

- Flamethrower
- Hardiness varies by species
- Partial shade
- Ample water

Named after a 19th century French naval commander, this genus of two species is native to the moist rainforests of New Caledonia, where it grows at lower elevations. They are highly ornamental, with single trunks, prominent crownshafts and crowns of pinnate leaves. They require warm, humid conditions.

C. lepidota. Hardy to 32°F/0°C. Single-trunked palm to 8 ft./2 m. tall with a 6 ft./2 m. spread. Green crownshaft. Leaves are stiffer and narrower than *C. macrocarpa.* Uncommon in cultivation.

C. macrocarpa. FLAMETHROWER, RED FEATHER PALM, RED LEAF PALM. Hardy to 28°F/-2°C. Slender, single-trunked palm to 24 ft./7 m. with a 16 ft./5 m. with a green crownshaft and a crown of dark green, curved pinnnate fronds. New fronds emerge bright red to maroon, fading to green as they age. Extremely attractive palm and an excellent landscape specimen for the tropics or warm subtropics. Needs well-drained soil. *C. 'hookeri'* is a variety with a cream-colored crownshaft. Also called *flame palm, houailou, blushing palm.*

clinostigma

- Hardiness to 40°F/4°C
- Partial shade
- Ample water

These pinnate palms from the South Pacific have prominent crownshafts and are highly ornamental. All require copious amounts of water and frost-free climates. Prefer acidic soils. Ideal palms for tropical areas, like Hawaii.

C. exorrhizum. Native to Fiji. Grows to 32 ft./11 m. with a spread of 12 ft./4 m. Looks similar to *Archontophoenix*, but more tropical.

C. gronophyllum. Native to Guadalcanal. Grows to 24 ft./7 m. with a spread of 12 ft./4 m. Stilt-rooted palm with stiff, green leaves.

C. harlandii. Native to Vanuatu. Grows to 24 ft./7 m. with a spread of 12 ft./3 m. Prominently ringed trunk and a lime green crownshaft.

C. ponapense. Native to the Caroline Islands. Tall palm to 64 ft./20 m. tall with a light gray trunk and a crown of arching pinnate leaves.

C. samoense. Native to Samoa. Tall palm to 60 ft./18 m. with a prominent pale green crownshaft and crown of finely divided fronds.

C. savoryanum. Native to the Bonin Islands. Slender palm to 24 ft./7 m. with a spread of 12 ft./4 m. Bears a prominently ringed trunk and a whitish crownshaft. Large crown of arching fronds.

Clinostigma samoense

coccothrinax

- Broom palms
- Hardiness varies by species
- Full sun
- Moderate water

From the Caribbean, Florida and Mexico, this genus of 49 species grows primarily in rocky, sunny areas. Most are indigenous to Cuba. Many are single-trunked fan palms and make attractive landscape specimens. Although they are slow growing, even young specimens are attractive. They require excellent drainage. All are relatively salt-tolerant and grow well in coastal areas. Require a tropical to subtropical climate. Very slow growing.

C. argentata. FLORIDA SILVER PALM, SILVER THATCH PALM. Hardy to 28°F/-2°C. Native to Florida, Mexico and the Caribbean. Slender-trunked palm grows slowly to 15 ft./5 m. tall with a spread of 6 ft./2 m. Deeply divided drooping leaves are dark green on the surface, silvery underneath. Very attractive as a landscape palm, able to withstand drought and harsh coastal conditions. Driving through the Florida Keys, you'll see them growing along the roadside. Also called *yuraguano de costa, Hispaniolan silver thatch palm.*

C. argentea. HISPANIOLAN SILVER PALM. Native to Hispaniola. Hardy to 28°F/-2°C. Similar to *C. argentata.*

C. barbadensis. SILVER THATCH PALM, SILVER PALM. Hardy to 27°F/-3°C. Native to Lesser Antilles, Puerto Rico and other Caribbean islands. Slender trunk to 20 ft./6 m. tall with a spread of 6 ft./2

Coccothrinax cupularis

Coccothrinax crinata

Coccothrinax borhidiana

Coccothrinax barbadensis

m. Similar to *C. argentata*, but leaves are larger. Also called *Lesser Antilles silver thatch, latanier, balai.*

C. borhidiana. Hardy to 28°F/-2°C. Native to Cuba. Single-trunked palm to 15 ft./ 5 m. tall with a spread of 10 ft./3 m. Stiff leaves.

C. crinita. OLD MAN THATCH PALM, THATCH PALM. Hardy to 28°F/-2°C., but can recover from 24°F/-4°C. Native to Cuba. Small single-trunked palm to 8 ft./2 m. tall with a spread of 10 ft./3 m. Trunk is covered with a dense, wooly fiber. Fronds have drooping leaves. Interesting conversation piece. Good container plant. Collector's item. Several subspecies are recognized, including short-haired old man palm (ssp. *brevicrinis*) and ssp. *crinita*, which has leaves with a shiny underside. Also called *guano barbudo, guano petate, mat palm.*

C. cupularis. Native to Cuba. May be a variant of *C. argentata.* Leaves are partially divided.

C. ekmanii. GOANE PALM. Native to Cuba, Haiti and Hispaniola. Slender single-trunked palm to 16 ft./5 m. tall. Leaves are stiff and look dry, colored tannish gray to silver. Excellent for seaside areas.

C. fragrans. FRAGRANT CUBAN THATCH PALM. From Cuba. Slender single-trunked palm to 26 ft./8 m. tall with a spread of 8 ft./2 m. Sparse crown of glossy fronds with a silvery underside. Fragrant yellow flowers.

C. hioramii. Native to Cuba. Medium-sized single-trunked palm that grows in sunny coastal areas, often in open sand.

C. inaguensis. THATCH PALM. Native to the Bahamas. Single-trunked palm to 16 ft./5 m. tall. Leaves on different specimens are variable; some are stiff, others having drooping leaflets. Grows in pure sand in its native habitat. Fragrant, creamy white flowers. Good coastal palm.

C. jamaicensis. JAMAICAN SILVER THATCH PALM. Hardy to 28°F/-2°C. Native to Jamaica. Single-trunked, clean-trunked palm to 26 ft./8 m. tall. Highly fragrant white flowers appear several times a year. In Jamaica, leaves are woven into hats and baskets. Can grow in pure sand and tolerates coastal salt spray.

C. miraguama. MIRAGUAMA PALM. Hardy to 25°F/-4°C. Native to Cuba. Single-trunked palm to 20 ft./6 m. tall with a spread of 10 ft./3 m. Trunk is covered with a wooly fiber; fronds are small, rigid and shiny green with a grayish underside. An attractive landscape palm.

C. munizii. Native to Cuba. Single-trunked fibrous trunk to 33 ft./10 m. tall. with a spread of 8 ft./2 m. Sparse, wispy crown. Rare.

C. proctorii. PROCTOR'S SILVER PALM, CAYMAN THATCH PALM. Hardy to 28°F/-2°C. Native to the Cayman Islands. Single-trunked palm to 26 ft./8 m. tall with a spread of 7 ft./2 m.. Smooth gray trunk with a crown of glossy green leaves. Creamy fragrant flowers. Leaves sometimes used to make rope. Tolerates coastal conditions.

C. spissa. SWOLLEN SILVER THATCH. Hardy to 30°F/-1°C. Native to the Dominican Republic and Haiti. Thick-trunked palm to 30 ft./9 m. tall with a spread of 15 ft./5 m. Sparse crown. Trunk becomes more swollen with age.

cocos nucifera

- Coconut palm
- Hardy to 33°F/1°C
- Full sun
- Ample water

Cocos nucifera

Nothing conjures up the image of the tropics like the coconut palm. Named for the Portuguese word for monkey. The origin is unknown, but coconut palms are found in nearly every tropical location, usually along sandy coasts. This genus has only one highly variable species. All are single-trunked palms up to 100 ft./30 m. with a spread up to 20 ft./6 m. Elegant crown of pinnate fronds and large, edible fruit. They require a tropical climate and cannot tolerate cold, wet winters. Coconut palms are highly tolerant of salt and heavy winds. Coconut palms are economically important for their fruit, fronds and wood. To grow a coconut palm, bury the seed halfway in moist sand or well-drained soil. Cultivars include:

dwarf golden Malay.' Abundant yellow fruit.
'Fiji dwarf.' One of the shortest cultivars.
'Jamaican tall.' Tall growing with a golden color to its leaves, fruit.
'Malayan semi-dwarf.' Abundant green, yellow or apricot fruits.
'Maypan.' Hybrid of 'Panama tall' and golden 'Malayan Dwarf."
'Panama tall.' Tall cultivar, moderately resistant to lethal yellowing.
'Samoan.' Developed in Samoa. Also fruits young.
'Spicata semi-dwarf.' Very tall with slimmer fronds. Red fruit.

copernicia

- Hardiness varies by species
- Full sun
- Moderate water

Copernicia baileyana

From the Caribbean, Florida and Mexico, this genus of 49 species grows primarily in rocky, sunny areas. Most are indigenous to Cuba. Many are single-trunked fan palms and make attractive landscape specimens. Although they are slow growing, even young specimens are attractive. They require full sun and excellent drainage. *Copernicia* are relatively salt tolerant and grow well in coastal areas. All require a tropical to subtropical climate.

C. alba. CARANDAY PALM, CARANDA PALM. Hardy to 23°F/-5°C. From Argentina, Bolivia, Brazil and Paraguay, this single-trunked palm with palmate fronds grows to 30 ft./9 m. with a spread of 6 ft./2 m. The stiff fronds are covered with wax, giving them a bluish color. Can recover from temperatures to 18°F/-8°C.

C. baileyana. BAILEY FAN PALM, BAILEY'S PALM. Hardy to 26°F/-3°C. From Cuba. This palm forms a massive light gray trunk to at least 40 ft./12 m. with a spread of 14 ft./4 m., larger in its native habitat. Trunk is topped by a dense, circular crown of fronds. Slow grower. Drought tolerant when mature. Also called *yarey, yarey hemba, yareyon.*

copernicia

Copernicia alba

Copernicia hospita

Copernicia macroglossa

Copernicia prunifera

C. glabrescens. GUANO PALM. Hardy to 18°F/-8°C. From Cuba, this clumping palm grows to about 25 ft./8 m. with a spread of 5 ft./2 m. Trunks are about 1 ft./.3 m. in diameter. More cold-tolerant than the others in this genus.

C. hospita. HOSPITA PALM. Similar to *C. glabrescens*, but highly variable in leaf color, ranging from green to silvery gray. Grows to 20 ft./6 m. tall with an 8 ft./2 m. spread. Single-trunked palm. Drought tolerant. Also called *guano cano, yarey.*

C. macroglossa. CUBAN PETTICOAT PALM, PETTICOAT PALM. Hardy to 30°F/-1°C. Native to Cuba, this palm grows to about 15 ft./3 m. tall with a similar spread. A highly ornamental palm that retains its fronds all the way to the ground, resembling a petticoat. Looks similar to a yucca. Drought tolerant. Also called *jata de guanabacoa, jata de guanbacoa.*

C. prunifera. CARNAUBA WAX PALM. Hardy to 30°F/-1°C. Native to Brazil. Single-trunked palm to 35 ft./11 m. tall with a spread of 10 ft./3 m. Leaves are glaucous blue-green in color. Wax from the leaves are the source of commercial car wax. Prefers ample water, but mature trees are drought tolerant.

corypha

Corypha utan

- Talipot palm
- Hardy to 32°F/0°C.
- Full sun
- Ample water

Native to India and a broad reaching area that includes the Philippines, Southeast Asia and northern Australia, this palm forms a massive trunk that grows to more than 70 ft./21 m. tall with a trunk diameter of more than 2 ft./1 m. Palmate leaves are nearly 20 ft./6 m. wide. This is a short-lived palm that produces the largest inflorescence in the plant world. Upon flowering, the tree will begin to decline and soon die. An impressive palm for large gardens in tropical areas. Slightly susceptible to lethal yellowing.

C. umbraculifera. TALIPOT PALM. Can reach 90 ft./30 m. tall. Not known in the wild.

C. utan. Smaller and less hardy than *C. umbraculifera.* Slightly susceptible to lethal yellowing.

cryosophila

- Rootspine palm
- Hardy to 26°F/-3°C
- Partial shade to full shade
- Ample water

Native to Mexico, Central America and northern Colombia, this genus of eight species grows in dry forests and rainforests at low to moderate elevations. Most are single-trunked, spiny palms with slender trunks and palmate fronds. Spines are actually sharp roots. Easy to grow and adaptable to poor, alkaline soils. Slow growing.

C. guagara. GUAGARA PALM. Native to Panama and Costa Rica. To 30 ft./9 m. Clumping or single-trunked palm with spiny roots covering the lower trunk.

C. stauracantha. Native to Nicaragua and Panama, this palm grows to 30 ft./10 m. tall with a 6 ft./2 m. spread.

C. warscewiczii. ROOTSPINE PALM. Native to Nicaragua, Costa Rica and Panama, this palm reaches 30 ft./9 m. tall with a 4 ft./1 m. spread. Leaves are lime green with a whitish underside. Also called *guáguara, palmas de escoba.*

C. williamsii. MOJARILLO PALM. Native to Honduras. Similar to *C. warscewiczii.* Rare in cultivation.

Cryosophila species

cyrtostachys renda

- Sealing wax palm, lipstick palm
- Hardy to 40°F/4°C.
- Full sun
- Ample water

Native to Borneo, Thailand and Malaysia, this clumping palm grows to about 35 ft./11 m. tall with a spread of about 20 ft./6 m. This palm gets its common name from its crownshaft, which ranges in color from vibrant, bright red to orange to deep yellow. This palm is extremely tropical and requires constant warmth and high humidity. This palm should be grown in soil amended with water-retaining organic matter. Also called *red sealing wax palm, maharajah palm, pinang palm.*

Cyrtostachys renda

dictyosperma album

Dictyosperma album

- Hurricane palm, princess palm
- Hardy to 29°F/-2°C.
- Partial shade to full sun
- Ample water

Native to islands in the Indian Ocean, this attractive single-trunked palm grows at a moderate pace to about 30 ft./9 m. tall with a spread of about 10 ft./3 m. Trunk diameter is about 6 inches/15 cm. with closely set rings. Attractive green crownshaft is topped by a sparse crown of pinnate fronds. Excellent for high wind areas or coastal locations. Needs well-draining soil.

dypsis

Dypsis decaryi

- Hardiness varies by species
- Partial shade to full sun
- Moderate water

Native to Madagascar, this genus typically grows as small understory to tall trees. Most make excellent container specimens; as indoor plants, they require ample light.

D. ambrositrae. Hardy to 28°F/-2°C. This palm grows as a single-trunked or multi-trunked tree to about 12 ft./4 m. tall with a 10 ft./3 m. spread. Has a sparse crown of fronds atop its 5 in./13 cm. trunks. Crownshaft is glaucous to gray green. Drought tolerant.

D. baronii. BARON PALM. Hardy to 28°F/-2°C, but can recover from temperatures to 19°F/-7°C. This clumping palm grows slowly to about 15 ft./5 m. tall with a spread of 6 ft./2 m. A very attractive palm with green ringed trunks and arching pinnate fronds. Needs good drainage.

D. cabadae. CABADA PALM. Hardy to about 32°F/0°C. This is an attractive, clumping palm that grows to about 20 ft./6 m. tall with a spread of more than 10 ft./3 m. Trunks have attractive whitish rings. Similar to *D. lutescens*, but trunks and crownshaft are green. Young palms are peach-colored.

dypsis

D. decaryi. TRIANGLE PALM. Hardy to 18°F/-8°C, this single-trunked palm grows to 30 ft./9 m. tall with a spread of 12 ft./4 m. Gray-green leaves are arranged in groups of three, giving the crown a triangle appearance. Sun and drought tolerant.

D. decipiens. MANAMBE PALM. Hardy to 22°F/-5.5°C. A single or multiple-trunked palm that grows slowly to 50 ft./15 m. tall with a spread of about 12 ft./4 m. Its large swollen trunk is topped with an open crown of arching, feathery leaves, sometimes tinged with blue.

D. lastelliana. RED-NECK PALM, TEDDY BEAR PALM. Hardy to 29°F/-2°C. Single-trunked palm to 60 ft./18 m. tall with a 15 ft./5 m. spread. Attractive whitish ringed trunk topped by a fuzzy, reddish crownshaft. Prefers full sun. Slow growing.

D. leptocheilos. TEDDY BEAR PALM. Hardy to 29°F/-2°C. Similar to *D. lastelliana*, but leaves are more droopy.

D. lutescens. ARECA PALM, BUTTERFLY PALM. Hardy to 25°F/-4°C. Common in cultivation, frequently seen as a houseplant. Slender yellow trunks grow to about 20 ft./6 m. tall with a 15 ft./5 m. spread. Also called *yellow bamboo palm, golden cane palm.*

D. madagascariensis. LUCUBENSIS PALM. Hardy to 29°F/-2°C. Clumping palm to 50 ft./15 m. tall. Ringed trunk and plumose fronds.

Dypsis lutescens

elaeis

- Oil palm
- Hardy to 32°F/0°C
- Full sun
- Moderate water

Next to the coconut palm, oil palms are the most important palms for commerce. Oils are extracted from the fruit and seed and used for industrial uses and as a food additive. As an avenue planting, these palms make an impressive statement with their large, drooping, pinnate fronds. Highly tropical in their requirements, these palms are frost sensitive and cannot tolerate dry air. They prefer moist soil that has been amended with organic matter.

E. guineensis. AFRICAN OIL PALM. Native to Africa, this palm is widely planted through the tropics for its palm and seed oil. A large tree that grows to at least 35 ft./11 m. with a 20 ft./6 m. spread, it looks best in a large area or as an avenue planting. Can tolerate soggy conditions and poor-draining soil. Also tolerates salty sprays in coastal areas. Also called *oil palm, macaw fat.*

E. oleifera. AMERICAN OIL PALM. Native to west forest areas of Central and South America. Smaller than the African oil palm, growing to about 15 ft./5 m. tall with a 12 ft./4 m. spread.

Elaeis guineensis

euterpe

Euterpe edulis

- Hardy to 25°F/-4°C
- Partial shade
- Ample water

One of the most beautiful palms from South America, these trees get their Latin name from the Greek goddess of song and poetry. Slender palms with graceful, arching pinnate fronds. Some have single trunks, others form moderate-sized clumps. All have prominent crownshafts which range in color from green to bright orange. These palms require a subtropical to tropical climate.

E. edulis. CABBAGE PALM, ASSAI PALM. Hardy to 25°F/-4°C. Native to Brazil, this single-trunked palm is grown throughout the world as an indoor specimen and as a landscape plant in subtropical to tropical areas. Grows to about 20 ft./6 m. tall with a 10 ft./3 m. spread. Also called *palmito palm, yayih, juncara palm.*

E. oleracea. ASSAI PALM. Hardy to 25°F/-4°C. This species is native to Brazil, Ecuador, French Guiana, Trinidad and Venezuela. This attractive clumping palm grows to about 40 ft./12 m. tall with a spread of about 10 ft./3 m. Slender stems are topped with prominent green crownshafts and beautiful feathery fronds.

gastrococos crispa

Gastrococos crispa

- Cuban belly palm, corojo
- Hardy to 32°F/0°C
- Full sun
- Ample water

Native to Cuba, this single-trunked palm grows moderately fast to 60 ft./18 m. tall with a 12 ft./4 m. spread. Fronds are spiny, plumose and blue-green in color. Trunk is light gray and swollen in the center. Younger plants bear spines on the trunk, which drop as the plant matures. This palm needs good drainage and plenty of water. An interesting palm because of its "belly-shaped" trunk. Suitable for tropical areas.

gaussia

- Hardiness varies by species
- Partial shade to full sun
- Moderate water

Native to Mexico, Central America and the Caribbean, this genus grows in dry, rocky regions in alkaline soils. These slow-growing palms do not exceed 20 ft./6 m. in height, making them ideal specimens for small yards. All are single-trunked palms with green crownshafts topped by a crown of pinnate fronds.

G. attenuata. PALM DE SIERRA. Hardy to 32°F/0°C. Native to Puerto Rico, this palm grows very slowly to 15 ft./5 m. tall with a 10 ft./3 m. spread. Trunk has a swollen appearance, similar to *Hyophorbe lagenicaulis.* Sparse crown of fronds.

G. maya. MAYA PALM, CAMBO. Hardy to 25°F/-4°C. Native to Mexico and Central America, this palm grows to about 18 ft./6 m. tall with a 10 ft./3 m. spread. Crown of fronds is sparse, usually only two or three at a time. Also called *palma cimaronna, palmasito.*

G. princeps. SIERRA PALM, PALMA DE SIERRA. Hardy to 32°F/0°C. Native to Cuba, this palm grows to about 15 ft./5 m. tall with a 10 ft./3 m. spread. Trunk is bottle-shaped, tapering at the crownshaft.

Gaussia maya

guihaia argyrata

- Guangxi palm, Chinese needle palm, dainty lady palm
- Hardy to 18°F/-7.8°C
- Partial shade
- Moderate water

Native to China and Vietnam. This clumping palm grows slowly on the limestone hills in Guangxi, China, where it reaches a height of 4 ft./1 m. tall with a spread of 6 ft./2 m. Trunks are covered with brown fibers and are topped with a crown of finely divided, fan-shaped leaves. Needs good drainage. Also called *Guilan dwarf palm.*

Guihaia argyrata

hedyscepe canterburyana

Hedyscepe canterburyana

- Umbrella palm, big mountain palm
- Hardy to 30°F/-1°C.
- Partial shade
- Moderate water

Native to Lord Howe Island in the South Pacific, this single-trunked palm is a great plant for containers or use as a specimen palm. Grows slowly to 35 ft./11 m. tall with a 15 ft./5 m. spread. Bears a prominent, silvery crownshaft topped by a dense crown of stiff, upright dark green fronds. It can tolerate shade and dry indoor conditions. Needs protection from sun as a young plant. Prefers moist, rich, well-drained soil.

heterospathe

Heterospathe elata

- Hardiness varies by species
- Shade to partial shade
- Ample water

This genus from Asia and the South Pacific contains several palms with ornamental qualities. All grow as understory trees, growing in the shade of larger trees. Some are single-trunked palms, others are clumping palms. Slender trunks are topped with pinnate fronds.

H. elata. SAGISI PALM. Hardy to 32°F/0°C. Native to the Philippines. This tall, slender single-trunked palm grows quickly to 40 ft./12 m. tall with a 20 ft./6 m. spread. Young fronds are pinkish to brownish, fading to green as they mature. Has become an invasive pest on Guam.

H. negrosensis. Hardy to 30°F/-1°C. Native to the Philippines. Similar to *H. elata*, but much shorter, reaching about 15 ft./5 m. in height with a spread of 6 ft./2 m. Good container specimen.

howea

- Kentia palm, sentry palm
- Hardy to 28°F/-2°C.
- Partial shade to full sun
- Moderate water

Native to Lord Howe Island in the South Pacific, these palms are commonly grown throughout the world as houseplants. Cultivated in Europe for nearly 150 years, these single-trunked palms grow slowly to 40 ft./12 m. with a spread of 10 ft./3 m. Can be grown in containers indefinitely, making them a great plant to overwinter indoors. With their elegant pinnate fronds, these palms are tropical in appearance. They can tolerate shade and dry indoor conditions. Excellent coastal plants. Provide protection from sun until several years old.

H. belmoreana. SENTRY PALM. Less common than *H. forsteriana* and perhaps more cold-sensitive. Leaves are lighter green and more erect than its relative. Also called *kentia palm.*

H. forsteriana. KENTIA PALM. One of the most common palms in cultivation, this tree makes a great outdoor specimen in subtropical to tropical climates. Looks good when planted in a grouping. Also called *sentry palm, thatch palm, forster sentry palm, thatch leaf palm.*

Howea forsteriana

hydriastele microspadix

- Hardy to 32°F/0°C
- Partial shade
- Ample water

Native to New Guinea, this clumping palm grows to 40 ft./12 m. tall with a 20 ft./6 m. spread. Tall, thin stems are topped with a sparse crown of pinnate fronds. An excellent specimen for subtropical to tropical climates.

Hydriastele microspadix

hyophorbe

Hyophorbe indica

Hyophorbe lagenicaulis

Hyophorbe verschaffeltii

- Hardiness varies by species
- Partial shade to full sun
- Ample water

Nearly extinct in their native habitat in the Mascarene Islands in the Indian Ocean, these palms are highly ornamental and ideal specimens for subtropical to tropical climates. All are single-trunked palms with prominent crownshafts and pinnate fronds. These palms are cold sensitive, although they can tolerate occasional frosts. They prefer well-drained soils.

H. indica. CHAMPAGNE BOTTLE PALM. Hardy to 30°F/-1°C. This palm grows quickly to 20 ft./6 m. tall with a 10 ft./3 m. spread. Crownshaft is a deep green color.

H. lagenicaulis. BOTTLE PALM. Hardy to 33°F/-1°C. Grows to 10 ft./3 m. tall with a 6 ft./2 m. spread. Prominently swollen trunk is bottle-shaped, hence its common name. Sparse crown of fronds. Ideal container specimen.

H. verschaffeltii. SPINDLE PALM. Hardy to 32°F/0°C. Grows to 15 ft./5 m. tall with a 6 ft./2 m. spread. Trunk is gray and slightly swollen. Beautiful blue-green crownshaft. Flowers are bright orange and extremely fragrant. Requires full sun.

hyphaene

Hyphaene dichotoma

- Hardiness varies by species
- Full sun
- Low water

Widely distributed in coastal areas from Africa to India, these palms are an oddity in the palm world: unlike most palms, these trees can form branching trunks. Large, stiff fronds are costapalmate-shaped with sharp thorns. Most are relatively drought-tolerant and can withstand occasional frosts. All love dry heat and full sun.

H. coriacea. ILALA PALM. Hardy to 28°F/-2°C. Native to Africa, this branching palm grows to 20 ft./6 m. tall with a 20 ft./6 m. spread. Leaves are blue-green. Also known as *H. natalensis.*

H. dichotoma. Hardy to 26°F/-3°C. Native to Western India. Similar to *H. coriacea,* but more massive. Grows to 60 ft./18 m. tall.

H. petersiana. VEGETABLE IVORY PALM. Hardy to 30°F/-1°C. Native to Africa, this single-trunked palm is similar to *H. coriacea*, but it rarely branches. Grows to 30 ft./9 m. tall with a 25 ft./8 m. spread.

H. thebaica. AFRICAN DOUM PALM, GINGERBREAD PALM. Hardy 29°F/-2°C. Native to Africa, this palm forms many branches and grows to 40 ft./12 m. tall with a 20 ft./6 m. spread. Grows enmasse along the Nile. This palm was important in ancient Egyptian times, often depicted in drawings. Preserved fruit has been recovered from the tombs of pharaohs. Fruit reportedly tastes like gingerbread.

johannesteijsmannia altifrons

- Joey palm, diamond joey, umbrella leaf palm
- Hardy to 30°F/-1°C.
- Full shade
- Ample water

Native to southeast Asia, these palms grow as understory plants in wet rainforests. A trunkless plant, its large fronds can reach more than 20 ft./16 m. tall with a 14 ft./4 m. spread. Leaves are palmate with a distinct wedge shape. Needs protection from sun, wind, frost and snails. A beautiful palm for tropical to warm subtropical climates.

Johannesteijsmannia altifrons

jubaea chilensis

- Chilean wine palm, coquito palm, honey palm
- Hardy to 4°F/-15°C
- Full sun
- Moderate water

Named for King Juba of Numidia in Africa, these palms are actually native to Chile, where they grow in dry valleys. These massive single-trunked palms grow slowly to 60 ft./18 m. tall with a spread of 25 ft./8 m. Trunks can reach a diameter of more than 3 ft./1 m. Feathery pinnate fronds are blue-green in color. The Chilean wine palm is endangered in its native habitat, where entire populations have been decimated by locals who fell the trees for its sugary sap. The sap is fermented, bottled and sold as wine or toddy in Chile. The Chilean government has restricted this practice and has launched a reforestation effort. Also called *palma chilena, palma de coquitos.*

Jubaea chilensis

jubaeopsis caffra

Jubaeopsis caffra

- Pondoland palm
- Hardy to 22°F/-6°C
- Full sun
- Ample water

Native to South Africa, this palm is related to the *Jubaea* genus and the coconut palm. This clumping palm with pinnate fronds grows slowly to 18 ft./6 m. tall with a 25 ft./8 m. spread. Sometimes looks unkempt because it retains its dead fronds.

kentiopsis oliviformis

Kentiopsis oliviformis

- Hardy to 30°F/-1°C
- Partial shade to full sun
- Moderate water

Native to New Caledonia, this palm looks similar to *Howea forsteriana*. This single-trunked palm with pinnate fronds grows slowly to 90 ft./27 m. tall with a 12 ft./4 m. spread. Trunk is greenish with prominent rings. Dark green fronds emerge from a pale green crownshaft.

laccospadix australasicus

- Atherton palm
- Hardy to 25°F/-4°C
- Partial shade
- Ample water

Native to the rainforests of Queensland, Australia, this rare palm is perfect for small gardens in California. This palm forms a clump of ringed trunks that are about 3 in./7 cm. in diameter. Grows to 12 ft./4 m. with a 10 ft./3 m. spread. Fronds are bright green. New growth is sometimes wine red in color. Fruit is an attractive bright red and looks like cherries. Good container specimen.

Laccospadix australasicus

latania

- Latan palm
- Hardy to 31°F/-1°C
- Partial to full sun
- Moderate water

The latan palm is native to the Mascarene Islands in the Indian Ocean, where it is rare because of deforestation. All are single-trunked fan palms, growing at a moderate pace to about 30 ft./9 m. tall with a 20 ft./6 m. spread. Stiff palmate leaves are blue-gray and large, often more than 12 ft./4 m. in length. Latan palms bear a slight resemblance to *Bismarckia nobilis.* Needs well-drained soil and a frost-free location. Species hybridize freely.

L. loddigesii. BLUE LATAN PALM. Native to Mauritius. Bears large gray leaves when mature. Young specimens have yellowish leaves with red veins.

L. lontaroides. RED LATAN PALM. Native to Reunion Island. Young plants have reddish leaf margins and veins. Mature trees are similar to *L. loddigessi.*

L. verschaffeltii. YELLOW LATAN PALM. Native to Rodrigues Island. Leaves more green than other species. Young leaves have yellow to orange leaf veins and petioles.

Latania loddigesii

lepidorrhachis mooreana

Lepidorrhachis mooreana

- Little mountain palm
- Hardy to 28°F/-2°C
- Partial shade
- Ample water

Native to Lord Howe Island, this palm grows to about 10 ft./3 m. tall with a 6 ft./2 m. spread. With its deeply recurved pinnate fronds, single trunk and diminutive stature, this palm makes an ideal small-scale specimen. Sensitive to dry winds and cold, this palm grows best in coastal areas where it can receive cool, humid air and well-drained soil. This palm is difficult to grow in hot or continually warm areas, including Florida and Southern California, but would make a fine palm for areas with climates similar to coastal San Francisco.

licuala

Licuala grandis

- Hardiness varies by species
- Shade
- Ample water

This large genus of fan palms consists of more than 100 species, ranging from Asia to Australia. Most are shade-loving smaller palms that live as understory plants in rainforests. Some are single-trunked, others form clumps of stems. Highly ornamental and excellent as foundation plantings. These palms are water lovers and are typically frost tender. All require excellent drainage. Most make good container specimens or houseplants.

L. cordata. Hardy to 32°F/0°C. Native to Sarawak, this small palm grows to about 6 ft./2 m. tall with a short trunk. Spreads to 5 ft./2 m. wide. Circular, undivided leaves are corrugated in appearance. Highly ornamental. A form exists with divided, wedge-shaped sections.

L. grandis. LICUALA PALM, VANUATU FAN PALM, RUFFLED FAN PALM. Hardy to 30°F/-1°C. Native to the Solomon Islands and Vanuatu. Forms a single trunk to about 10 ft./3 m. tall. topped by a full crown of undivided, circular leaves. Spreads to 6 ft./2 m. wide. Widely grown in tropical areas, this palm also makes a great container plant, but needs ample humidity and continually moist soil. Needs shade when young, filtered sun as a mature tree.

licuala

L. orbicularis. Hardy to 32°F/0°C. Native to Sarawak. A highly ornamental palm with shiny, round undivided leaves. Compact growth habit to 8 ft./2 m. makes it an excellent container specimen. Spreads to 4 ft./1 m. wide. Similar to *L cordata,* but can be distinguished by slightly wedge-shaped, less circular fronds. Needs ample humidity and warmth.

L . peltata. ELEGANT PALM. Hardy to 28°F/-2°C. Native to Malaysia, India, Bangladesh and Thailand. Grows as a single-trunked palm to about 20 ft./6 m. tall with a 5 ft./2 m. spread. Slender trunks are topped with a full, round crown of leaves. Two forms exist: 'elegans' has divided fronds, 'sumawongii' has undivided fronds.

L. ramsayi. AUSTRALIAN FAN PALM. Hardy to 26°F/-3°C. Native to Queensland, Australia. A single-trunked palm that can grow to 60 ft./18 m. tall, but usually grows to 30 ft./9 m. in cultivation with an 18 ft./6 m. spread. Deeply divided leaves are wedge-shaped.

L. spinosa. SPINY LICUALA PALM, MANGROVE FAN PALM. Hardy to 26°F/-3°C. Native to a vast range in Southeast Asia. Forms a dense clump of divided leaves to about 10 ft./3 m. tall with a similar spread. More cold hardy than most. Can be used as a hedge or container plant.

Licuala ramsayi

linospadix monostachya

- Walking stick palm
- Hardy to 23°F/-5°C
- Partial shade
- Ample water

Native to Australia, these single-trunked palms with pinnate fronds grow to 8 ft./2 m. tall with a spread of 3 ft./1 m. Because of their size and light requirements, these palms make great indoor container specimens that can be moved outdoors during the warmer months. This palm gets its common name from its use during World War I. The stems of this species were harvested by the thousands and carved into walking sticks for returning wounded soldiers.

Linospadix monostachya

livistona

Livistona chinensis, saribus and *decora*

Livistona australis

Livistona muelleri

Livistona benthamii

Livistona chinensis

- Hardiness varies by species
- Full sun
- Moderate water

Named for Patrick Murray, Baron of Livingston, this genus of about 28 species ranges from northern Africa through Asia and Australia, where most species are native. They resemble *Washingtonia* but are typically more elegant, with shiny palmate fronds. All make great container plants but, as indoor plants, require strong light. Frond stalks of younger trees are armed with very large, curved spines, making them unsuitable around children.

L. australis. AUSTRALIAN FAN PALM. Hardy to 10°F/-12°C. Native to Australia, this palm grows to 50 ft./15 m. with a spread of 9 ft./3 m. Also called *Australian cabbage tree palm, Gippsland palm, fan palm.*

L. benthamii. Hardy to 12°F/-11°C. Native to Australia and New Guinea. Grows quickly to 50 ft./15 m. tall with a spread of 8 ft./2 m. Attractive tan trunk is covered with knobby leaf scars

L. chinensis. CHINESE FAN PALM. Hardy to about 12°F/-11°C. Native to Taiwan and Japan, these palms grow to 30 ft. tall with a spread of 10 ft./3 m. Bright green, shiny leaves droop at the segment tips.

L. decora. RIBBON FAN PALM. Hardy to 12°F/-11°C. Native to Australia, this palm grows relatively quickly to about 60 ft./18 m. tall with a spread of 7 ft./2 m. With its finely divided, weeping fronds, this is one of the most attractive palms in the genus.

L. drudei. Hardy to about 24°F/-4°C. Native to Australia, where it grows slowly to 50 ft./15 m. tall with a spread of 7 ft./2 m. Bears a large open crown of stiff and erect divided fan leaves.

L. inermis. WISPY FAN PALM. Hardy to 24°F/-4°C, this Australian native grows to 20 ft./6 m. tall with a spread of about 8 ft./2 m. A slender palm with a sparse crown of wispy gray-green fronds.

L. jenkinsiana. MOUNTAIN FAN PALM. Hardy to 21°F/-6°C. Native to Thailand, Malaysia and Myanmar, where it grows slowly to about 40 ft./12 m. tall with a spread of 15 ft./5 m. Bears a crown of large, divided leaves that are green above and blue-green on the underside.

L. mariae. CENTRAL AUSTRALIAN CABBAGE PALM. Hardy to 25°F/-4°C. From Central Australia, where it grows slowly to about 30 ft./9 m. tall with a spread of 10 ft./3 m. Young plants often have a reddish appearance. Also called *Central Australian fan palm.*

L. muelleri. DWARF FAN PALM. Hardy to 22°F/-6°C. This palm is native to Australia and New Guinea, where it grows in grassland areas. Slow grower to 20 ft./6 m. with a spread of 7 ft./2 m.

L. saribus. TARAW PALM, SERDANG. Hardy to 18°F/-8°C. Native to Southeast Asia, Indonesia and Philippines. Grows slowly to about 20 ft./6 m. tall with a 10 ft./3 m. spread; in its native habitat, it will reach 80 ft./24 m. Bears large clusters of brilliant blue fruit. Prefers partial shade, especially when young.

lodoicea maldivica

- Double coconut
- Hardy to 35°F/2°C
- Full sun
- Ample water

Native to the Seychelles Islands, this palm bears the largest seed in the plant world. A beautiful fan palm that is rarely seen outside of botanical or collector's gardens, this tree is slow growing, sometimes taking more than two decades to form a trunk, eventually reaching more than 80 ft./24 m. tall. with a 20 ft./6 m. spread. Palms are grown from seed, which cost several thousand dollars each. Seeds should be planted in the palm's permanent location; upon germination, the seed will send a long sinker root deep into the soil, making it almost impossible to transplant. Germination can take more than a year. Needs a tropical climate with ample warmth, humidity and well-drained soil to survive.

Lodoicea maldivica

lytocaryum

- Hardy to 25°F/-3°C
- Partial shade
- Ample water

From southeastern Brazil, these small palms grow as understory plants in the rainforest. All are single-trunked palms with slender trunks and arching pinnate fronds. Good container specimens.

L. hoehnei. An attractive small palm with finely pinnate fronds that grows slowly to about 10 ft./3 m. tall with a spread of about 5 ft./2 m. Undersides of leaflets are white.

L. weddellianum. VICTORIAN PARLOR PALM, MINIATURE COCONUT PALM. Growing to 12 ft./4 m. tall with a 5 ft./2 m. spread, this palm makes an elegant container plant and is ideally suited for indoor use. Dark green, shiny fronds. Also called *Weddell palm, feather palm, ague, ica.*

Lytocaryum weddellianum

medemia argun

Medemia argun (juvenile)

- Hardy to 28°F/-3°C
- Full sun
- Low water

Native to the Sudan, this extremely rare palm was considered extinct until it was recently rediscovered growing in Wadi Delah in Sudan. An attractive fan palm with blue-green fronds. Grows at a moderate rate to about 30 ft./9 m. tall with a 20 ft./6 m. spread. Looks similar to *Washingtonia* species. Extremely tolerant of drought, severe winds, salinity and sun. Not readily available in retail stores; seek specialty nurseries or obtain seed from a commercial palm seed supplier. Related to *Bismarckia* and *Hyphaene* genera.

metroxylon

Metroxylon warburgii

- Sagu palm
- Hardy to 35°F/2°C
- Partial shade to full sun
- Ample water

Native to the South Pacific, these palms are grown primarily for thatch and sago, a staple part of the diet of millions of people in Asia. All make attractive landscape specimens with upright, pinnate fronds. Some are single-trunked palms, others are clumping palms. They are best suited for humid, wet tropical areas. Except for *M. amicarum*, these palms die after flowering.

M. amicarum. CAROLINE IVORY NUT PALM. Native to Micronesia. Single-trunked palm that grows to 66 ft./20 m. tall with a spread of 34 ft./10 m. Spiny leaves. The kernels of the seeds are used as vegetable ivory, suitable for carving and making buttons.

M. sagu. SAGO PALM. Native to New Guinea. Grown primarily for the sago that is used as a food source. A clumping palm that grows quickly to 66 ft./20 m. tall.

M. vitiense. Native to Fiji. Single-trunked palm with a large trunk that grows to 49 ft./15 m. tall with a 33 ft./10 m. spread.

M. warburgii. Native to Vanuatu and Western Samoa. Grows quickly to 23 ft./7 m. tall with a 33 ft./10 m. spread. Crown is particularly erect.

nannorrhops ritchiana

- Mazari palm
- Hardy to -5°F/-21°C
- Full sun
- Moderate to little water

Native to the Middle East, Pakistan and Afghanistan, these clumping palms grow to 8 ft./2 m. with a spread of 10 ft./3 m. The trunk of this palm is primarily subterranean, providing it with protection from extreme temperatures and firestorms. Older specimens may sport branching trunks. Several forms are now readily available in cultivation with costapalmate fronds that range from blue-green to silver gray. Very drought, wind and cold tolerant. Needs well-drained soil.

Nannorrhops ritchiana

neoveitchia storckii

- Unleito
- Hardy to 32°F/0°C
- Full sun
- Ample water

Native to Fiji, this attractive single-trunked palm grows at a moderate to fast pace to 40 ft./12 m. tall with a 20 ft./6 m. spread. Bears a heavy, full crown of long pinnate fronds. In its native habitat, these trees are felled and used to support pole buildings. Needs ample humidity. Tolerant of various soils.

Neoveitchia storckii

normanbya normanbyi

Normanbya normanbyi

- Black palm, Queensland black palm
- Hardy to 25°F/-4°C
- Partial shade
- Ample water

Native to the rainforests of northern Queensland, Australia, this single-trunked palm grows slowly to 15 ft./5 m. tall with a 10 ft./3 m. spread. Bears a 6 in./15 cm. thick trunk that is topped with a light green crownshaft and head of plumose fronds. An excellent small landscape palm for tropical to subtropical areas.

nypa fruticans

Nypa fruticans

- Mangrove palm
- Hardy to 32°F/0°C
- Full sun
- Ample water

Native to Southeast Asia, these clumping palms grow to 12 ft./4 m. with a similar spread. This palm is an aquatic species that grows in brackish waters and coastal mangroves in its native habitat. Trunks are subterranean. This palm is an important economic plant in its native habitat, used for sugar, vinegar, alcohol and thatch. A novelty palm that is rare in landscape cultivation, but popular in botanical gardens. This palm would make an interesting focal point for wet coastal areas in tropical climates.

oncosperma tigillarium

- Nibung palm
- Hardy to 32°F/0°C
- Partial shade to full sun
- Ample water

Native to Southeast Asia, this clumping palm grows to 60 ft./18 m. tall with a 20 ft./6 m. spread. Individual stems are 6 in./15 cm. thick and are covered with black spines, which limit its use as a landscape palm. Needs a large lot; not suitable for smaller properties. Fronds are bright green and pinnate. An attractive palm for wet, tropical areas, able to thrive in poor-draining swampy areas. Highly salt tolerant.

Oncosperma tigillarium

parajubaea

- Hardiness varies by species
- Full sun
- Moderate water

Native to South America, these tall single-trunked palms are highly ornamental and tropical in appearance. Pinnate fronds are graceful and elegant. Trunks on young trees are covered with fiber and resemble coconut palms.

P. cocoides. QUITO PALM. Hardy to 22°F/-5.6°C, this palm is unknown in the wild and may be a cultivated species. Grows to 60 ft./18 m. with a spread of 16 ft./5 m. Hardy and reasonably fast growing, this palm is becoming popular in cultivation because of its resemblance to more tropical species. Also called *quito coconut palm, mountain coconut, coco cumbe, cumbe palm, coco.*

P. torallyi. JANCHICOCO PALM, PALMERA ZUNCA. Native to Ecuador, this palm is hardy to 24°F/-4°C. Grows to 40 ft./12 m. tall with an 18 ft./6 m. spread. Leaflets are blue green and more finely divided than *P. cocoides.* In their native habitat, these palms are subject to high winds and extreme drought.

Parajubaea cocoides

phoenix

Phoenix canariensis

Phoenix reclinata

- Date palm
- Hardiness varies by species
- Full sun
- Moderate water

Native to Africa, Crete, the Middle East, Asia, the Philippines, Indonesia and the Canary Islands, the *Phoenix* genus is among the most important in the palm world. Many are used as ornamental landscape plants. The streets and parks of subtropical regions like the Riviera and Southern California are characterized by *P. canariensis. P. dactylifera* has commercial significance as the source of edible dates; dates from *P. sylvestris* are also sold commercially and used for palm sugar. All have feathery, pinnate fronds and sharp spines. These palms hybridize easily, so make sure seed comes from a source that can guarantee a pure species. Many in this genus seem to require ample magnesium, so a regular fertilizer routine is essential to maintaining their health and green fronds.

P. canariensis. CANARY ISLAND DATE PALM, PINEAPPLE PALM, CANARY DATE PALM. Hardy to 20°F/-7°C. This massive single-trunked palm from the Canary Islands grows relatively quickly to 70 ft./21 m. tall with a spread of 30 ft./ 9m. Thick crown of deep green fronds. This is the familiar palm that lines the streets of Beverly Hills and the towns of the Riviera. Severely pruned young plants look like large pineapples. Drought and wind tolerant.

P. dactylifera. EDIBLE DATE PALM, DATE PALM. Hardy to 15°F/-9°C, this is the date palm of commerce. Native to the Middle East. Grows to 80 ft./24 m. tall with a spread of 15 ft./5 m. Trunk more slender and canopy more sparse than *P. canariensis*; fronds are more gray-green. This tree naturally suckers to form a clump of trunks. Very wind and drought tolerant. Only fruits reliably in dry climates. Cultivars include 'Deglet Noor', 'Medjool' and Zahedi.'

P. loureiroi. VOIAVOI PALM, LOUREIR'S DATE PALM, DWARF DATE PALM. Hardy to 22°F/-6°C, this palm is native to India, Indochina and Hong Kong. Grows to 8 ft./2 m. tall with a spread of 10 ft./3 m. Single-trunked and clumping forms exist. Maintains its persistent leaf bases along the trunk. Similar to *P. roebelinii.*

P. pusilla. CEYLON DATE PALM. Hardy to 22°F/-6°C. From India this single-trunked palm grows slowly to about 9 ft./3 m. tall with a spread of 10 ft./3 m. Trunk has persistent leaf bases that point upward, topped by a crown of bright green, feathery fronds.

P. reclinata. SENEGAL DATE PALM, AFRICAN WILD DATE PALM. Hardy to at least 20°F/-7°C, this clumping palm from Africa grows quickly to 30 ft./9 m. with a spread of more than 25 ft./8 m. Slender, arching trunks are topped with a crown of drooping green fronds, giving this palm a distinctly tropical appearance. Offshoots can be removed to form a single-trunked tree. Looks best with regular maintenance to thin suckering offshoots. Needs ample water.

P. roebelenii. PYGMY DATE PALM. Hardy to 22°F/-6°C, this dwarf palm grows to 20 ft./6 m. tall with a 6 ft./2 m. spread. From Laos, this palm typically grows as a single-trunked specimen, but some clump into multiple-trunked specimens. This is an ideal palm to create a tropical mood and a perfect landscape specimen. Fronds are shiny green and graceful; young trees are sometime confused with ferns. These trees grow slowly, taking 20 years or more to reach full size. Great container specimen, but requires bright direct light indoors. Also called *dwarf date palm, miniature date palm, roebelenii palm.*

P. rupicola. CLIFF DATE PALM. Hardy to 24°F/-4°C, this small palm grows on rocky cliffs and gorges in its native India. Forms a single trunk to 25 ft./8 m. tall with a spread of 15 ft./5 m. Similar to *P. roebelenii,* but taller. Less spiny than other *Phoenix* and perhaps the most tropical in appearance. In some ways, resembles a small coconut palm. A variegated form exists. Also called *wild date palm, India date palm, Indian wine palm.*

P. sylvestris. SILVER DATE PALM. WILD DATE PALM, SUGAR DATE PALM. Hardy to 22°F/-6°C. From India, Nepal, Myanmar and Pakistan. This beautiful single-trunked palm grows to about 40 ft./12 m. tall with a 20 ft./6 m. spread. Similar to *P. canariensis*, but more refined with gray-green fronds. In India, this palm's fruit is sold as edible dates, and flower stalks are tapped for their sap, which is used to make palm sugar. Also called *toddy palm, India date palm.*

P. theophrasti. CRETAN DATE PALM. Hardy to 10°F/-12.2°C. From Greece, Ceylon and Turkey, this clumping palm grows to about 50 ft./15 m. tall with a spread of more than 20 ft./6 m. Looks similar to *P. dactylifera.*

Phoenix roebelenii

Phoenix sylvestris

Phoenix rupicola

Phoenix theophrasti

Phoenix loureiroi

pinanga

Pinanga caesia

Pinanga coronata

- Hardiness varies
- Partial shade to full sun
- Ample water

A large genus of more than 100 species ranging from India through Southeast Asia and China, these small palms with pinnate fronds are ideally suited for the landscape. Most are small understory palms that grow in the shade of taller trees. Some grow as single-trunked palms, others grows as clumping plants. Well-suited for tropical areas with high humidity and ample moisture, although some can grow in dry, subtropical areas. Some species have highly ornamental mottled leaves. Most make excellent container specimens.

P. caesia. Hardy to 32°F/0°C. Native to Sulawesi. Single-trunked palm grows to 12 ft./4 m. tall. Trunk is about 3 in./8 cm. in diameter and is topped with an attractive orange-colored crownshaft, turning purple toward the top. New fronds are bronze, fading to blue-green.

P. coronata. CORONATA PALM, IVORY CANE PALM. Hardy to 29°F/-2°C. Native to Java and Sumatra, this clumping palm grows slowly to 10 ft./3 m. tall with an 8 ft./2 m. spread. Individual stems are 2 in./5 cm. diameter. Stems are green and bamboo-like topped with an ivory crownshaft.

P. javana. Hardy to 22°F/-6°C. Native to Java, this single-trunked palm grows slowly to 12 ft./4 m. tall with an 8 ft./2 m. spread. Trunk is topped by a brilliant green crownshaft and an attractive crown of dark green, pinnate fronds. Bright red fruit is an added attraction.

P. philippinensis. Hardy to 30°F/-1°C. Native to the Philippines, this clumping palm grows to 10 ft./3 m. tall with a 5 ft./2 m. spread. Looks similar to a clumping betel nut palm (*Areca*) with ringed 1 in./3 cm. stems, swollen crownshaft and erect fronds. Excellent container specimen.

pritchardia

- Loulou palm
- Hardiness varies by species
- Partial shade to full sun
- Moderate water

Native to the islands of the Pacific, this genus includes the only true native palms on the Hawaiian Islands, where they are called loulou palms. These single-trunked fan palms prefer tropical climates with temperatures above 50°F/10°C, although some Hawaiian species can be grown in warm, subtropical areas.

P. affinis. KONA FAN PALM. Hardy to 28°F/-2°C. Native to the Big Island of Hawaii. This palm grows quickly to 30 ft./9 m. tall with an 18 ft./6 m. spread. Leaves are bright green.

P. arecina. Hardy to 32°F/0°C. Native to the Hawaiian Island of Maui. Visitors to Maui's Hana Coast can see native palms growing on the hillside. Grows to 39 ft./12 m. tall with an 18 ft./6 m. spread.

P. beccariana. Hardy to 30°F/-1°C. Native to the Big Island of Hawaii. Grows to 40 ft./12 m. tall with a 14 ft./4 m. spread. Leaves are large and nearly round.

P. hillebrandii. Hardy to 30°F/-1°C. Native to the Hawaiian Island of Molokai. Grows slowly to 25 ft./8 m. tall with a 12 ft./4 m. spread. A form with blue-green fronds exists.

P. kaalae. Hardy to 30°F/-1°C. Native to the Hawaiian Island of Oahu. Grows to 25 ft./8 m. tall with a 12 ft./4 m. spread. Thick, deep green fronds have drooping segment tips.

P. lanaiensis. Hardy to 30°F/-1°C. Native to Lanai, Hawaii, where it grows on the steep, dry cliffs. Leaves are thin and covered with brown scales. Grows to 25 ft./8 m. tall with a 12 ft./4 m. spread.

P. lowreyana. Hardy to 30°F/-1°C. Native to the Hawaiian Island of Molokai. Grows to 12 ft./4 m. tall with a 6 ft./2 m. spread.

P. martii. DWARF HAWAIIAN FAN PALM. Hardy to 30°F/-1°C. Native to the Hawaiian Island of Oahu. This small palm grows to 9 ft./3 m. tall with a 7 ft./2 m. spread. Large leaves are blue-green.

P. pacifica. FIJI FAN PALM, PACIFIC FAN PALM. Hardy to 40°F/4°C. From Tonga, this palm grows to 25 ft./8 m. tall with a 12 ft./4 m. spread. Bears a dense crown of bright green fronds. Highly tropical in its requirements. Also called *Fijian fan palm.*

P. remota. NIHOA PALM. Hardy to 28°F/-2°C. Native to the remote uninhabited Hawaiian Island of Nihoa. Grows at a moderate pace to 18 ft./6 m. tall with a 6 ft./2 m. spread. Bright green, waxy leaves. More wind tolerant than most.

P. thurstonii. THURSTON PALM. Hardy to 40°F/4°C. From Fiji. Similar to *P. pacifica*, but taller, reaching up to 50 ft./16 m. tall with a 12 ft./4 m. spread. Leaves roundish or diamond-shaped and light green.

Pritchardia thurstonii

Pritchardia arecina

Pritchardia lowreyana

Pritchardia pacifica

Pritchardia remota

pseudophoenix

Pseudophoenix vinifera

- Hardy to 26°F/-3°C
- Full sun
- Low water

Native to Florida, Mexico, Belize and the Caribbean, these single-trunked palms with pinnate fronds grow in open coastal areas or dry hills with alkaline soils. Trunks grow to about 20 ft./6 m. tall with a 12 ft./4 m. spread and are characterized by prominent rings. All have swollen crownshafts and a sparse crown of feathery fronds. These trees are attractive landscape palms and are well suited for coastal areas in subtropical to tropical climates. Highly salt tolerant, these palms need well-drained soil. Slow growing.

P. sargentii. CHERRY PALM, BUCCANEER PALM. Most commonly grown of the species. Green trunk with white rings is topped by a prominent crownshaft. Leaves are blue-green. Bright red fruit. Also called *palma de guinea, cacheo, kuka*.

P. vinifera. BUCCANEER PALM, WINE PALM. Native to the Dominican Republic and Haiti. Similar to *P. sargentii*, but with a thicker trunk. Wine is distilled from the sweet juice of its fruit. Also called *cherry palm, cacheo, katie*.

ptychosperma

Ptychosperma elegans

- Hardy to 29°F/-2°C
- Partial shade
- Ample water

Native to northern Australia, New Guinea and the South Pacific, this genus of about 28 species grows in a range of habitats, including rainforests and wet swamplands. Some are single-trunked palms, others form clumps. Slow growing with slender trunks, these pinnate palms make excellent landscape plants for subtropical to tropical climates. All thrive with high humidity and require good drainage.

P. elegans. SOLITAIRE PALM, ALEXANDER PALM. Native to Australia, this single-trunked palm grows to 60 ft./18 m. tall with a 10 ft./3 m. spread. Trunk is about 6 in./15 cm. diameter. One of the best in the genus for subtropical climates. Ideal for landscaping or as a large container specimen.

P. macarthurii. MACARTHUR PALM. Native to Australia, this palm forms a clump to 25 ft./8 m. tall with a 7 ft./2 m. spread. Trunks more slender than *P. elegans*, typically 3 in./8 cm. diameter. Marginal in subtropical areas. Best suited for warmer, tropical climates. Excellent indoor specimen for bright, well-lit areas.

raphia

- Raffia palm
- Hardiness varies by species
- Partial shade to full sun
- Ample water

This genus of about 28 species is native to Africa, Madagascar and tropical America. Single-trunked or clumping palms, all have huge pinnate fronds that are used to produce raffia twine. In Africa, the palms are of significant importance as a source of building materials and food items, including wine which is produced from fermented fruit. As garden plants, these palms are only suited for large properties.

R. farinifera. RAFFIA PALM. Native to East Africa and Madagascar. Hardy to 29°F/-2°C. This clumping palm grows to at least 40 ft./12 m. tall with a 30 ft./9 m. spread. Each trunk is monocarpic and will die after flowering, but new suckers will fill in. Needs ample humidity to look good.

R. hookeri. RAFFIA PALM. Native to western Africa. Similar to *R. farinifera*, but some varieties sport single trunks.

Raphia farinifera

ravenea

- Hardiness varies by species
- Partial shade to full sun
- Ample water

This genus is native to Madagascar and the Comoros Islands. Single-trunked palms are topped with an attractive crown of pinnate leaves.

R. glauca. Hardy to 28°F/-2°C. Native to Madagascar, this palm grows to 15 ft./5 m. with a 10 ft./3 m. spread. Similar to *R. rivularis*, but much smaller. Leaves are blue-green and finely divided. Prefers partial shade.

R. hildebrandtii. Hardy to 28°F/-2°C. Native to the Comoros Islands near Madagascar. Small palm growing slowly to 10 ft./3 m. tall with a 7 ft./2 m. spread. Showy leaves. Endangered in its natural habitat. Good small palm for warm temperate to tropical climates. Needs partial shade and well-drained soil.

R. rivularis. MAJESTY PALM, MAJESTIC PALM. Hardy to 26°F/-3°C. Native to Madagascar, this palm has become popular in cultivation because of its rapid growth. Grows to 40 ft./12 m. tall with a 12 ft./4 m. spread. Needs copious water and frequent applications of fertilizer. Good container specimen for outdoor patios. Also called *bakaly, gora, malio.*

Ravenea rivularis

rhapidophyllum hystrix

Rhapidophyllum hystrix

- Needle palm, porcupine palm, spine palm
- Hardy to -20°F/-29°C
- Full sun
- Moderate water

Native to the southeast United States, this palm is one of the most cold hardy, able to sustain temperatures well below freezing. Grows slowly to about 6 ft./2 m. tall with a spread of 10 ft./3 m. This palm with palmate fronds forms a clump of dense fronds armed with long, sharp spines on the leaf sheaths and upper trunks. Will grow in shade, but prefers sun. In its native habitat, this palm typically grows in damp or swampy areas; provide moderate water in cultivation. Also called *blue palmetto, creeping palmetto, dwarf saw palmetto, spine palm, vegetable porcupine, hedgehog palm.*

rhapis

Rhapis excelsa

- Lady palm
- Hardiness varies by species
- Partial shade
- Ample water

Attractive small clumping palms with palmate fronds that make great container and indoor specimens. All form clumps of slender, fiber-covered trunks. Individual trunks are sometimes used to craft walking sticks and canes. Somewhat frost tender but can be overwintered indoors and moved outside during the warmer months.

R. excelsa. LADY PALM. Hardy to 18°F/-8°C. Native to Japan and China, this palm grows slowly to 12 ft./4 m. tall with a 6 ft./2 m. spread. Can survive indefinitely in a container. Tolerates extreme shade, but prefers filtered light. Variegated forms exist. Also called *bamboo palm, ground rattan, fern rhapis, miniature fan palm.*

R. humilis. SLENDER LADY PALM. Hardy to 21°F/-6°C. Native to China. Similar to *R. excelsa*, but taller, more airy. Leaflets are longer and more pointed. Grows to about 18 ft./5 m. tall with a spread of more than 6 ft./2 m. Also called *reed rhapis.*

R. subtilis. DWARF LADY PALM. Hardy to 24°F/-4°C. Native to Southeast Asia, this palm grows to about 4 ft./1 m. tall and wide. Smaller than *R. excelsa* with fewer leaf segments. Also called *Thailand lady palm.*

rhopaloblaste elegans

- Hardy to 32°F/0°C
- Partial shade
- Ample water

Native to Guadalcanal. This slender palm grows to 39 ft./12 m. tall with a 12 ft./4 m. spread. Sports a pinkish crownshaft. Fronds are narrow and pinnate and form an elegant crown. This palm grows as an understory palm.

Rhopaloblaste elegans

rhopalostylis

- Hardy to 24°F/-4°C
- Partial shade to full sun
- Ample water

From the South Pacific, these palms all have slender gray trunks with prominent crownshafts. Fronds are pinnate and bright green. All need protection from sun until established and all thrive in moist, well-drained soils. Protect from damaging winds.

R. baueri. NORFOLK ISLAND PALM. From Norfolk and Kermadec Islands, this palm grows slowly to 40 ft./12 m. tall with a spread of about 15 ft./5 m.

R. cheesemanii. KERMADEC ISLAND PALM. From Raul Island, this palm grows to about 50 ft./15 m. tall with a spread of 15 ft./5 m. Similar to *R. baueri.*

R. sapida. NIKAU PALM, SHAVING BRUSH PALM. From Norfolk Island and other areas of New Zealand, this palm has a bulging green crownshaft and upright fronds that give it the appearance of a large brush once commonly used to apply shaving cream. Smaller than *R. baueri*, growing to about 30 ft./9 m. tall with a 10 ft./3 m. spread. This palm was heavily used by native Maoris as a source of food and thatch for roofing. Also called *feather duster palm.*

Rhopalostylis baueri

roystonea

Roystonea regia (L) and Roystonea oleracea (R)

Roystonea borinquena

Roystonea princeps

Roystoneas at Fairchild Tropical Botanic Garden

- Royal palm
- Hardiness varies by species
- Full sun
- Ample water

Named for General Roy Stone, an officer in the United States Army who served in Puerto Rico, these palms are native to the Americas, including Florida, the Caribbean and Central and South America. Royal palms are among the most majestic and useful as landscaping plants. Any visitor to Miami, Florida has seen these tall palms on the city's avenues and driveways, their gray trunks looking like towers of concrete lining the streets. All are single-trunked palms with bright emerald green crownshafts topped by a crown of pinnate fronds. Highly tolerant of strong winds, these palms have adapted to hurricane conditions by dropping their fronds in severe storms. They require a subtropical to tropical climate. All need good drainage and perform poorly in cold soil. All make outstanding landscape palms.

R. altissima. JAMAICAN ROYAL PALM, MOUNTAIN CABBAGE ROYAL. Native to Jamaica, these palms are hardy to 30°F/-1°C. These palms grow quickly to 40 ft./12 m. tall with a 16 ft./5 m. spread. Leaves droop below the crownshaft.

R. borinquena. PUERTO RICAN ROYAL PALM. Native to Puerto Rico, Haiti, the Virgin Islands and the Dominican Republic, this palm is hardy to 30°F/-1°C. Grows moderately fast to 35 ft./11 m. with a 15 ft./5 m. spread. Similar to *R. regia*, but shorter and with more plumose fronds. Also called *palma caruta, palma real, palmis, palmiste.*

R. oleracea. SOUTH AMERICA ROYAL PALM, VENEZUELAN ROYAL PALM. Native to Colombia, Trinidad, Tobago, Barbados and Venezuela, this palm is hardy to 32°F/0°C. Grows to 150 ft./46 m. tall with a 25 ft./8 m. spread. The tallest of the species, this palm is easily identified by its upright crown of fronds which do not droop below a horizontal plane, giving them a more upright appearance. Also called *cabbage palm, royal palm, mapora.*

R. princeps. JAMAICAN ROYAL PALM. Hardy to 32°F/0°C. Native to Jamaica, this palm has a thin trunk and grows quickly to 40 ft./12 m. with a 15 ft./5 m. spread.

R. regia. CUBAN ROYAL PALM, FLORIDA ROYAL PALM. Hardy to 27°F/-2°C. Native to Cuba, Belize, Honduras, Cayman Islands, Mexico and possibly Florida. This is one of the most commonly planted in the genus, growing moderately fast to 60 ft./18 m. tall with a 16 ft./5 m. spread. Considered by many to be one of the best landscaping palms. The Florida royal palm, referred to as *R. elata*, is a synonym. Also called *royal palm, yagua, palma criolla, palma real.*

sabal

- Palmetto
- Hardiness varies by species
- Full sun
- Moderate water

Native to the Americas and the Caribbean, these slow-growing fan palms are highly adaptable and extremely hardy. Some grow to more than 50 ft./15 m. tall and resemble *Washingtonia* palms, others have subterranean trunks and look like small shrubs. The edible buds of many are harvested as palm hearts; this practice has led to the decimation of several species. All are drought, wind and salt tolerant, and some are among the most cold tolerant in the palm family. Difficult to transplant until a trunk forms.

S. bermudana. BERMUDA PALM, BERMUDA PALMETTO. Hardy to 15°F/-9.4°C. Native to Bermuda, this single-trunked palm grows to 20 ft./6 m. tall with a spread of 12 ft./4 m. Thick gray trunks are topped by a crown of large bluish-green leaves.

S. causiarum. PUERTO RICAN HAT PALM, PANAMA HAT PALM. Hardy to 14°F/-10°C. Native to the Caribbean, this palm forms a massive, single trunk up to 30 ft./9 m. tall with a 16 ft./5 m. spread. Trunks can reach 4 ft./1 m. in diameter. In Puerto Rico, fronds are used to weave hats. Leaves are usually dull green. Also called *palma cana, palma de sombrero, yarey, hat palm.*

S. domingensis. Hardy to 22°F/-6°C. Native to Cuba and Hispaniola. Similar to *S. causiarum*, but leaves are more blue-green. Also called *palma cana, latanier-chapeau.*

S. etonia. SCRUB PALMETTO. Hardy to 5°F/-9°C. This palm is native to Florida and bears a subterranean trunk, rarely reaching to 8 ft./2 m. in height. Leaves are yellow-green. Relatively uncommon in cultivation.

S. mexicana. OAXACA PALMETTO, TEXAS PALMETTO. Hardy to -5°F/-15°C. Native to Texas, Mexico and Central America, this single-trunked palm reaches 50 ft./15 m. tall with a spread of 12 ft./4 m. In El Salvador, plantations are grown for thatch. A robust tree with a heavy gray trunk and a large crown of green fronds. Very drought tolerant. Also called *Rio Grande palmetto, Texas sabal, palma de micharo, soyate, palma de sombrero.*

S. minor. DWARF BLUE PALMETTO, SCRUB PALMETTO. Hardy to -5°F/-15°C, this palm is among the most cold hardy. Native to the southeastern United States, the dwarf palmetto rarely reaches 6 ft./2 m. tall with a 6 ft./2 m. spread. Its subterranean trunk helps insulate this palm from volatile climates. Fronds are bluish-green. Needs moderate water and will grow in partial shade to sun. Also called *dwarf palmetto, little blue stem, swamp palmetto, bush palmetto, latanier.*

Sabal palmetto

Sabal minor

sabal

Sabal causiarum

Sabal domingensis

Sabal mexicana

Sabal leaf bases

S. palmetto. PALMETTO PALM, CABBAGE PALM. Hardy to 1°F/-17°C. Native to the Bahamas, Cuba and southeastern United States, this single-trunked palm grows to 20 ft./6 m. tall with a 10 ft./3 m. spread. Trunks are latticed with the frond bases, or boots, adding to their appeal. Leaves are dark green to yellow-green. Visitors to the Florida Everglades can see large stands of these trees in their native habitat. Also called *sabal palm, blue palmetto, cabbage palmetto, cabbage tree, common palmetto, guana cana, palma cana.*

S. rosei. SAVANNAH PALMETTO. Hardy to 15°F/-9°C. From Mexico, this palm grows to 60 ft./18 m. tall with a spread of 14 ft./4 m. Similar to *S. palmetto*, but taller and more slender. Also called *palma de llano.*

S. uresana. SAVANNAH PALMETTO. Hardy to 18°F/-8°C, this single-trunked palm from Mexico grows to 30 ft./9 m. tall with a 14 ft./4 m. spread. Trees have a large crown of glaucous leaves. Common in western Mexico, from sea-level to more than 4,000 ft./1,500 m. Also called *palma blanca.*

satakentia liukiuensis

Satakentia liukiuensis

- Satake palm
- Hardy to 32°F/0°C
- Partial shade to full sun
- Ample water

From the Ryuku Islands of Japan, this single-trunked palm with pinnate fronds grows to 30 ft./9 m. tall with a 10 ft./3 m. spread. Trunk is less than 6 in./15 cm. diameter. Large crown of dark green fronds on a purplish crownshaft. New, emerging leaves are sometimes pinkish-purple. Looks slightly like a coconut palm, but trunks are usually more green with distinct rings. Somewhat wind tolerant. Excellent for tropical areas, but can be grown in subtropical areas if planted in a good microclimate with protection from frost. Prefers rich, moist soil.

schippia concolor

- Silver pimento palm, mountain pimento palm
- Hardy to 30°F/-1°C
- Partial shade to full sun
- Moderate water

From Belize and Guatemala, this single-trunked palm with palmate fronds grows slowly to 15 ft./5 m. tall with a 5 ft./2 m. spread. Trunk is about 3 in./8 cm. diameter. Leaves are dark green and very glossy. Looks similar to *Coccothrinax* and *Thrinax* palms. Excellent palm that will grow in warm temperate areas if given winter protection, especially when young. Needs excellent drainage.

Schippia concolor

serenoa repens

- Saw palmetto
- Hardy to -5°F/-21°C
- Full sun
- Moderate water

From the southeastern United States, this palm is extremely cold hardy, able to sustain prolonged periods in sub-zero weather. Multiple trunks grow slowly to 6 ft./2 m. tall with a spread of more than 10 ft./3 m. Trunks are primarily subterranean. In its native habitat, saw palmetto forms a dense groundcover under native pine trees and on sand dunes along coastal areas. Fruit is used as a dietary supplement to prevent prostate cancer. Several forms are grown, with fronds ranging from green to blue-green to silver-blue in color.

Serenoa repens (blue form)

syagrus

Syagrus romanzoffiana

Syagrus amara

Syagrus cearensis

Syagrus romanzoffiana

Syagrus sancona

- Hardiness varies by species
- Full sun
- Moderate water

This large genus of palms includes several species which are relatively cold hardy. Native to the Americas, these palms can tolerate relatively dry conditions. All have pinnate fronds and most have single trunks. Several species are grown commercially for palm oil. The *Syagrus* genus is the closest relative of the coconut palm and many produce edible seed that tastes like coconut. Suitable for warm, temperate to tropical climates. All require good drainage.

S. amara. OVERTOP PALM. Native to the West Indies, this palm is hardy to 31°F/-1°C. Grows to 40 ft./12 m. tall with a 20 ft./6 m. spread. Looks similar to a coconut palm, with a slender single trunk and dark green pinnate fronds. Relatively drought and salt tolerant. Also called *yattahou, coco nain, petit coco de bois, petit, mocho.*

S. cearensis. Hardy to 32°F/0°C. Native to Brazil. A clumping palm which grows moderately quickly to 30 ft./9 m. tall. Individual trunks are about 6 in./16 cm. diameter. An attractive clumping palm. Relatively drought tolerant.

S. comosa. BABAO, CATOLE. Native to Brazil, this palm is hardy to 30°F/-1°C. A medium-sized palm growing up to 20 ft./6 m. tall with a 10 ft./3 m. spread, this tree is ideal for small gardens. Typically a single-trunked tree with a 3 in./8 cm. thick trunk, this palm can also clump with multiple stems. Fruit is edible.

S. coronata. LICURY PALM. Native to Brazil, this palm is hardy to 27°F/-2°C. These single-trunked palms grow slowly to 12 ft./4 m. tall with a 10 ft./3 m. spread. Persistent leaf bases form a lattice on the trunk. Relatively drought tolerant. Also called *Licuri palm, ouricuri, nicury, Licuryseiro.*

S. pseudococos. COCO VERDE. Native to Brazil. Hardy to 31°F/-1°C, this single-trunked palm bears a resemblance to the coconut palm and grows to 30 ft./9 m. tall with a 16 ft./5 m. spread. Also called *coco amargoso, piririma.*

S. romanzoffiana. QUEEN PALM. Hardy to 22°F/-6°C. Native to Brazil, Argentina, Paraguay and Uruguay, this single-trunked tree is one of the most common palms in cultivation. Grows quickly to 65 ft./20 m. tall with a spread to 20 ft./6 m. Pinnate fronds are plumose. Also called *Cocos plumosa, Arecastrum, chiriva, jeriba, pindo.*

S. sancona. Native to South America, this tree grows to 50 ft./15 m. tall with a 12 ft./4 m. spread. Hardy to 25°F/-4°C. This solitary palm has a relatively slender trunk. Also called *sumuque, tiba, sarare.*

S. schizophylla. ARICURY PALM. Native to Brazil, this palm is hardy to 28°F/-2°C. A small, single-trunked palm that grows slowly to 10 ft./3 m. tall, this tree bears long, persistent leaf bases that form a lattice on the trunk. Fronds are finely divided and dark green. Makes a good outdoor container specimen. Also called *Aricuriroba, licurioba.*

thrinax

- Hardiness varies by species
- Partial shade to full sun
- Moderate water

From the Americas, this genus grows in alkaline soils primarily in coastal areas. All are small, slender, single-trunked fan palm that are slow-growing and require excellent drainage. Excellent medium-sized tree for coastal areas.

T. excelsa. BROAD THATCH PALM. Native to Jamaica, this palm is hardy to 28°F/-2°C and grows to 14 ft./4 m. tall with a 9 ft./3 m. spread. Leaves larger than other species with a whitish underside.

T. morrisii. BRITTLE THATCH PALM. Hardy to 27°F/-3°C. Native to Florida and the Caribbean, this palm grows to 30 ft./9 m. tall with a 9 ft./3 m. spread. Leaves sometimes blue-green.

T. parviflora. BROOM PALM. Hardy to 22°F/-6°C. Native to Jamaica, this palm grows to 12 ft./4 m. tall with a spread of 6 ft./2 m. Does not tolerate much cold when young. A subspecies has heavily textured leaves that appear corrugated.

T. radiata. FLORIDA THATCH PALM. Hardy to 27°F/-3°C. Native to Florida, the Caribbean, Mexico and Belize. Grows to 16 ft./5 m. tall with a 9 ft./3 m. spread. Will tolerate more shade than others. Needs excellent drainage and will tolerate alkaline soils.

Thrinax radiata

trachycarpus

- Windmill palm
- Hardiness varies by species
- Full sun
- Moderate water

Among the most cold hardy groups of palms, the *Trachycarpus* genus includes eight species of single-trunked palms from northern India, Myanmar, Nepal, northern Thailand and China. These palmate-leaved palms bear small crowns of fronds on trunks that are covered with coarse fibers. In their native habitats, these palms grow in full sun at high elevations in mountainous areas. Excellent container specimens that can live in large tubs indefinitely.

T. fortunei. CHINESE WINDMILL PALM, CHUSAN PALM. Hardy to 0°F/-18°C. Native to China, this palm grows to 30 ft./9 m. tall with a spread of 5 ft./2 m. Makes an ideal palm for cold climates; also does well as an indoor palm if given bright, indirect light. Protect from winds, which will damage its fronds. Fibers that cover its trunk are used in China to make brooms, clothing and doormats. Also called *windmill palm, hemp palm.*

T. latisectus. WINDEMERE PALM. A rare palm from India, this species is similar to *T. fortunei* and considered equally cold hardy.

Trachycarpus fortunei

trachycarpus

Trachycarpus martianus

T. martianus. HIMALAYAN WINDMILL PALM. Hardy to 5°F/-15°C. Native to India, Nepal and Myanmar, this palms grows slowly to 20 ft./6 m. tall with a spread of 6 ft./2 m. Trunk is gray with darker rings. Less cold hardy than *T. fortunei.* Also called *khasia palm, Martius windmill palm, Martius fan palm.*

T. nanus. YUNNAN DWARF PALM. Hardy to 7°F/-14°C. From western China, this palm grows to about 2 ft./1 m. tall with a spread of 6 ft./2 m. Fronds are deep green to silver-blue. Rare in cultivation, but seed is now readily available.

T. takil. TAKIL PALM, KUMAON FAN PALM. Hardy to -3°F/-19°C. This species from the Himalayas in India grows at a moderately fast pace to 40 ft./12 m. tall with a spread of 6 ft./2 m. Similar to *T. fortunei*, but taller and with larger fronds. May be the most cold hardy in the *Trachycarpus* genus. Tolerates snow and frost. Most palms sold as *T. takil* are actually *T. fortunei.*

trithrinax

Trithrinax brasiliensis

- Hardiness varies by species
- Full sun
- Moderate water

From South America, these three species are native to cold, dry areas. All are drought and cold tolerant and require good drainage. Trunks are covered with sharp spines and topped with a dense crown of palmate leaves.

T. brasiliensis. SPINY FIBER PALM, BRAZILIAN NEEDLE PALM. Hardy to 15°F/-10°C. Native to Brazil, this palm grows slowly to 20 ft./6 m. tall with a spread of about 10 ft./3 m. Plants may be clumping or have a single trunk. Leaves are finely divided and deep green to gray-green. Trunk is covered with a dense fiber and spines. Widely adapted to alkaline soils. Can rot in shady, water-logged areas. Also called *buriti, carandai, saho palm, Saro.*

T. campestris. CORONDAY PALM. Hardy to 13°F/-11°C. From Uruguay and Argentina, this clumping palm grows slowly to 15 ft./5 m. tall with spread of more than 10 ft./3 m. Leaf color is gray-green to silver-blue. Very picturesque palm. Also called *Saro, caranday.*

veitchia

- Hardiness varies by species
- Partial shade to full sun
- Ample water

Native to the tropical rainforests of Fiji, Vanuatu and the Philippines, these elegant single-trunked trees have slender gray trunks topped by a prominent crownshaft and deep to bright green pinnate fronds. All need tropical to warm subtropical climates. Excellent landscape palms for the right climate, although some are highly susceptible to lethal yellowing disease. Young plants make excellent container specimens for bright indoor locations.

V. arecina. MONTGOMERY PALM, SUNSHINE PALM. Hardy to 29°F/-2°C. Native to Vanuatu, this palm grows moderately fast to 30 ft./9 m. tall with a 10 ft./3 m. spread. Bears a ringed trunk and greenish-white crownshaft. Syn. *Veitchia montgomeryana.* Also called *arecina palm.*

V. filifera. Native to Fiji, this palm is hardy to 32°F/0°C. Grows quickly to 30 ft./9 m. tall. Needs a tropical climate.

V. joannis. JOANNIS PALM. From the Philippines and Fiji, this palm is hardy to 31°F/-1°C. Highly ornamental with a bright green crownshaft and ringed trunk. Grows quickly to more than 60 ft./18 m. tall.

V. winin. WININ PALM. Native to Vanuatu, this palm is hardy to 32°F/0°C and grows to 49 ft./15m. tall with a 15 ft./5 m. spread. Bears a pale green crownshaft and stiff, arching dark green fronds.

Veitchia arecina

verschaffeltia splendida

- Seychelles stilt palm
- Hardy to 40°F/4°C
- Full sun
- Ample water

Native to the Seychelles, this highly ornamental palm grows as a single-trunked specimen to 30 ft./9 m. tall with a spread of 10 ft./3 m. Aerial roots which rise out of the ground give it a stilt-like appearance. Leaves are pinnate, but are undivided on younger specimens. Trunks are covered with small black spines. Needs a tropical climate and warm greenhouse environment. Also called *stilt palm, splendid stilt palm.*

Verschaffeltia splendida

washingtonia

Washingtonia filifera (L) and *Washingtonia robusta* (R)

- Hardiness varies by species
- Full sun
- Moderate water

Native to California and Mexico, these single-trunked fan palms grow quickly to more than 60 ft./18 m. tall, making them best suited for streets and avenues. A familiar sight in Los Angeles, California where they line the motorways and avenues of Santa Monica, Hollywood and Beverly Hills. Sometimes called petticoat palms because of the downturned dead fronds that persist on their trunks. In cultivation, these fronds are usually trimmed off.

W. filifera. CALIFORNIA FAN PALM, PETTICOAT PALM. Hardy to 15°F/-9°C. The only palm native to California and Arizona, this palm grows quickly to 60 ft./18 m. tall with a spread of 14 ft./4 m. Light green fronds form an open crown. Trunk is stout and shorter than *W. robusta*. Very wind and drought resistant. Also called *desert fan palm, California palm, California cotton palm, northern Washingtonia.*

W. robusta. MEXICAN FAN PALM, SKYDUSTER. Hardy to 22°F/-5°C. Native to Mexico, this palm grows quickly to 100 ft./30 m. with a spread of 12 ft./4 m. Palm is topped by a dense crown of medium green fronds. Looks like *W. filifera*, but taller, thinner and with shorter-stemmed fronds. Also called *Washington palm, Mexican Washingtonia, thread palm, southern Washingtonia, Washingtonia.*

welfia regia

Welfia regia (juvenile)

- Palmito, amargo
- Hardy to 32°F/0°C
- Full sun
- Ample water

Native to Central and South America, this single-trunked palm grows to 60 ft./18 m. tall with a 20 ft./6 m. spread. Bears a crown of dark green pinnate fronds. Emerging fronds are a striking red color. Needs shade when young, but is sun tolerant as it ages. Requires ample water and rich, fertile soil. Cannot tolerate frost. Also called *palma conga, camara.*

wodyetia bifurcata

- Foxtail palm
- Hardy to 30°F/-1°C
- Full sun
- Moderate water

Native to Australia, this palm was unknown to botanists until the 1980s; just a few years ago, these trees were considered a collector's item. This single-trunked palm grows quickly to 30 ft./9 m. tall with an 8 ft./2 m. spread. Pale green crownshaft and ringed trunk. Leaves are pinnate with a plumose appearance. Prefers loose, well-draining soil. A perfect avenue planting for small residences. Often called a "mini royal palm" because of its passing resemblance to the *Roystonea* genus.

Wodyetia bifurcata

zombia antillarum

- Zombie palm, zombi
- Hardiness to 30°F/-1°C
- Partial shade
- Moderate water

Native to Dominican Republic and Haiti, this clumping palm grows slowly to 6 ft./2 m. tall with an 8 ft./2 m. spread. Bears sharp downward-pointing spines. Leaves are palmate. Similar to *Coccothrinax* and *Thrinax*, this palm makes an excellent landscape plant if situated where its spines are not a nuisance. Good for subtropical to tropical climates. Also called *latayne pikan, latayne zombi, guanito.*

Zombia antillarum

Caring for your palm

Spiny palms

Many of the most beautiful palms bear spines on their trunks and leaves. Avoid planting spiny palms in areas that are frequented by children and pets, including walkways or other outdoor recreational areas, like swimming pools. Fortunately, the spines of many palms, like *Phoenix*, become less of a nuisance as the palms mature and spines grow out of reach.

Coconut palms *(Cocos nucifera)* tolerate strong winds, high salt levels and occasional dry spells, making them ideal for coastal areas in the wet or dry tropics. Their falling fruit, however, creates a maintenance problem and can cause serious injury.

Picking the right palm

Coconut palms are beautiful, but their falling nuts can be deadly. And the spines of a date palm can cause serious injury. More than one palm lover has chosen the wrong palm for their location. Here's a quick checklist to help you choose the right palm:

☐ **Size and spread.** A small one-gallon palm may grow to more than 80 ft./24 m. tall with a 20 ft./6 m. spread. Make sure you know the ultimate height of the palm you are planting. And, maybe more important, understand the spread. A palm may only reach 20 ft./6 m. tall, but it may eventually take up more than 400 square feet as it forms a large clump. The good news is that palms are very predictable in their growth; you simply need to plan for their ultimate size.

☐ **Light.** Make sure you know the light requirements of the palm you are planting and match its needs with your location. Also, consider future sun availability; if you have planted other trees in the area, will they eventually block the light?

☐ **Water needs.** If you are planning a Mediterranean garden, choose a drought-tolerant species, like *Chamaerops humilis*. If you are planting amongst tropicals, choose a water-loving tree, like *Chambeyronia macrocarpa*. Many desert palms will suffer in locations that receive too much water or humidity.

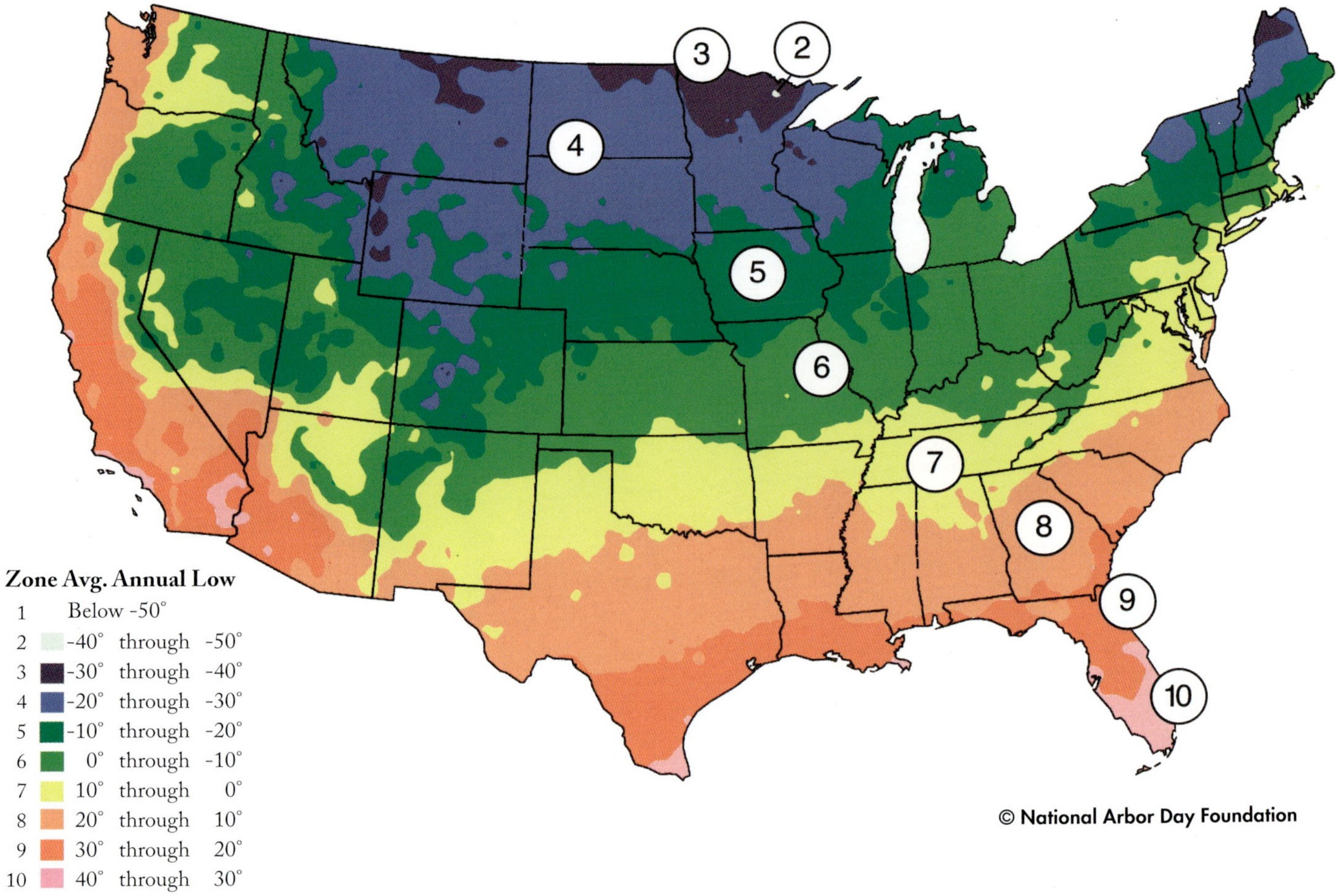

☐ **Temperature.** A surprising large number of palms are very cold tolerant, able to withstand temperatures below freezing. If you live in a cold location, choose a cold-hardy palm unless you are willing to take precautions to winterize your palm. Similarly, cool weather palms, like *Ceroxylon alpinum*, will decline in hot, humid climates, like Hawaii and Florida. The cold tolerance data in this guide are based on the latest research, but the numbers are inexact. Your microclimate, including humidity level, and the length of the cold spell will greatly influence the cold-hardiness of palms in your garden. Many palms can endure exposure for short periods, measured in days, not weeks. Use these numbers as a good starting point.

☐ **Trunk type.** Do you want to frame a view or block one? If you want to frame a vista, choose a tall, single-trunked palm, like *Roystonea regia* or *Syagrus romanzoffiana.* And to block an unsightly view, choose a clumping species, like *Rhapis excelsa*.

☐ **Leaf type.** Really a matter of preference, palms with feathery leaves seem to evoke the tropics more than palms with fan-shaped leaves. (Interestingly, the only native palms to Hawaii are fan palms.) Most palm gardens mix fan palms with feather palms to add visual interest.

☐ **Growth rate.** Like cold-hardiness, determining growth rate is complicated. Soil, temperature, fertilizer and other variables greatly affect the growth rate. Tropical species will grow much faster in Hawaii than they do in California or Florida. Use this guide as a good starting point.

☐ **Salinity tolerance.** Typically, palms that are native to coastal areas are more salt-tolerant than palms that are native to the tropics. Palms absorb salt from soil, the water they receive and from the air, usually in the form of sea spray. If you live in a coastal area, pay particular attention to the salinity tolerance of the species you plant and be prepared to wash the palm's foliage with fresh water.

☐ **Spines.** Some of the most beautiful palms, like the stilt palm (*Verschaffeltii splendida*), are armed with spines. Do you have children or pets? Or is your location near a walkway or pool? Make sure you consider the dangers of planting spiny palms.

☐ **Indoor potential.** There is no such thing as an indoor palm. Palms grow outdoors, but we can simulate outdoor growing conditions by moderating temperature, humidity, light and moisture. Some palms are surprisingly adaptable to indoor growing.

The Santa Barbara courthouse in California provides these palms with added warmth and wind protection.

Palms as hedges?

You usually think of palms as tall, towering trees, but many make excellent hedges. If you need to screen an unsightly object or you would like to add privacy to your yard, consider planting a hedge of palms, like this lady palm *(Rhapis excelsa)*. Look for palms that form clumps, rather than single-trunked varieties. If you need something taller, the Senegal date palm (*Phoenix reclinata*) will screen out multi-story buildings.

Choosing the right location for your palm

Good planning will ensure you get years of enjoyment from your palm with a minimum of fuss

The location you choose for your palm garden will determine your ultimate success in growing palms. Several primary factors must be considered when choosing a good location.

Sun. Make sure you understand the light requirements for the palms you are planting. Many palms, like *Brahea armata*, will suffer if planted in the shade. Likewise, most *Chamaedorea* palms will burn if exposed to afternoon sun. In nature, sun-loving trees create a canopy for shade lovers. Plant fast-growing trees, including *Caryota* and *Archontophoenix* species, to create a shady canopy for shade-loving palms.

Soil. Although some palms can tolerate extreme soils, like coconut palms in sand, most prefer well-drained soil that contains organic material. Good soils allow air and water to penetrate them easily. Clay soils retain water and may prevent your palm's roots from receiving much-needed air. Sandy soils may drain too quickly, preventing your palm trees from absorbing water and nutrients. Some soils should be amended with peat moss or decomposed plant material. Clay soils can also be amended with pumice or course sand. Only use washed sand that does not contain salt. A good soil mix should hold together, but crumple apart easily in your hands.

Microclimate. It's important to understand the concept of microclimates, particularly if you are going to grow palm trees in frost-prone areas. A microclimate is a localized climatic area that

Microclimates are a key to growing palms successfully, regardless of your location.

differs from the surrounding areas. A city, for example, may be several degrees warmer than surrounding rural areas. Concrete, asphalt and radiant heat from buildings all affect the temperature in a city or town. Conversely, a large undeveloped expanse of land will be significantly colder because of its exposure to the elements. Coconut palms typically do not grow in California. Yet, if you visit a small parking lot near the ocean in Newport Beach, you will see a mature coconut palm growing quite well. The reflected heat and protection of the surrounding buildings, coupled with the heat-retaining black pavement, have dramatically altered the air and soil temperature around the trees.

On a smaller scale, your neighborhood—and even your garden—may contain several microclimates, each differing by several degrees in the various seasons. The southwest side of your property is probably the warmest, but an enclosed courtyard on the north side may ultimately be a better location.

Wind. Most palms are fairly tolerant of wind, but dry winds can dessicate the fronds of many tropical species. Plant palms that are native to desert areas, like the blue hesper palm (*Brahea armata*) or provide protection near a structure or a windbreak of trees. In general, feather palms tolerate wind better than fan palms.

Salt. Most coastal palms can tolerate high levels of salinity in the air and soil; palms that are native to inland areas will usually show various levels of frond damage. If you live in an area with coastal breezes or saline soil, plant salt-tolerant palms like *Serenoa repens* or *Phoenix dactylifera*.

Understanding highs and lows

To pick the best location, you must understand the microclimates in your garden. You'll need a few thermometers:

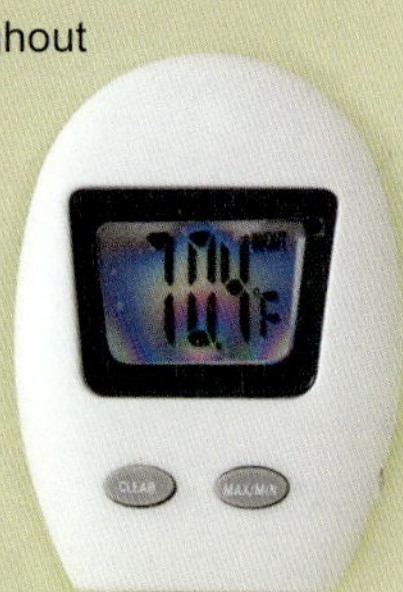

1 Place several thermometers throughout your garden. Minimum-maximum thermometers store the highest and lowest temperatures they record. Attach your thermometers to wooden stakes to keep them about 24 in./61 cm. off the ground. Choose locations close to buildings, walls, windbreaks, streams and ponds. High elevations and southern locations are usually good locations for palm trees. Cold air will flow down a hill, making low spots susceptible to colder temperatures.

2 Read your thermometers in the morning. Write down the temperatures for each thermometer to determine the warmest location.

Seven ways to improve the microclimate for your palms

Here are a few factors that will affect the microclimate in your garden:

1 Elevation. Hot air rises and cold air falls. If your house is on a hill, the coldest air will drain down the slope away from your property. During the winter months, a hillside is usually warmer than the valley below. If your house is on the downside of a slope, however, cold air will form a pocket next to the wall facing the slope.

2 Buildings and structures. Buildings lose heat, often enough to keep the foundation zone several degrees warmer than the rest of the property. Buildings also reflect heat from the sun and block cold, desiccating winds. On a frosty morning, sheltered areas will remain several degrees warmer and be devoid of frost.

3 Walls. Stone, block or brick will reflect and retain heat. A location next to a well-situated wall may provide enough heat and wind protection to increase the temperature, and create a heat sink, by several degrees.

4 Sun exposure. The south side of your house or apartment will nearly always be the warmest area in your yard. The northwestern side will usually be the coldest, unless this area is protected from the weather.

5 Paved surfaces. Paved surfaces retain heat. An asphalt driveway will remain warm for several hours after the sun goes down.

6 Water. Water helps regulate the climate around it. In the winter, a location near a pool, pond or creek will often remain several degrees warmer than surrounding areas.

7 Surrounding vegetation. Hedges and tall evergreens, like oaks and conifers, provide ideal protection from the harsh weather. Consider planting additional trees to surround your palm garden.

The growing conditions in nurseries and greenhouses vary greatly from your garden. Humidity and light are the primary differences.

Why do palms sulk?

Palm growers will tell you that palms tend to sulk after they are purchased.

They are describing a phenomenon that is sometimes referred to as "greenhouse shrink." Unless they are field grown in conditions similar to your garden, palms will suffer a certain amount of setback, regardless of your expert care.

Typically, palms planted in the ground will decline slightly before they fully acclimate. Fronds may wilt and brown, giving the palm a ragged appearance. Root systems may take many months to establish, and new frond growth may take longer.

Many gardeners have impulsively removed palms that appear ragged after planting, but palm growers know it takes time for a palm to adjust to its new surroundings.

Growers recommend you wait a year to judge your palm's condition. By then, your palm will be fully established and flourishing in its new environment.

Acclimating your palm

Sun, size and season are the factors that will ensure your palm gets off to a good start

Proper acclimation will help ensure your success with your new palm tree. Any changes to the environment will be a shock to the plant's systems. Temperature, light, wind and humidity will differ from the growing conditions your palm has experienced indoors. The primary factors to consider include age and sun exposure.

What is the best size palm to plant?

Small seedlings are usually too tender to plant in the garden. If you have small palm seedlings, plant them in two-gallon containers. After the roots have filled the container, your palm is probably ready to be planted.

Don't rush; wait until the palm has a strong root system. This typically takes at least a year or longer.

The larger the palm, the faster it will adapt, particularly if you are planting a species that is marginally cold tolerant for your area.

Surround small palms with a wire fence to deter rabbits, deer and other animals from destroying the plants.

Age. Palm trees become hardier with age. Young fleshy seedlings will succumb to inclement conditions more easily than mature trees. In order for a palm to survive harsh climates that are different from its normal growing conditions, the plant must have a fully developed root system and a woody trunk. After you plant your tree outdoors, your palm will take several years to fully establish itself.

Size. The larger the palm and the more established its root system, the better. Optimally, your palm should be grown to at least a two-gallon size before you plant it permanently in the garden. The palm should be relatively root-bound with its roots occupying most of the container. Plan to overwinter your young palms indoors until they reach this size.

Sun. Palms that are grown indoors or in partial shade must be gradually exposed to full sun. Place your palm outdoors and cover it with a light sheet. Remove the sheet at the end of the day to allow the palm to receive full sun for an hour. Every other day, increase the amount of time the palm is uncovered until the palm no longer requires a cover.

Plant in spring or early summer. The best time to plant is when the soil is warm. Spring or early summer is typically best because your palm will not be stressed by the extreme heat, hot winds and drought that typically accompany later summer. If you live in a desert climate, avoid planting during the hottest months of the year. In South Florida, Hawaii and coastal Southern California, however, the planting season extends nearly year round, provided you supply adequate water.

If you are planting a palm that has been grown indoors, make sure night temperatures do not dip below 55°F/13°C before you leave your palm outside overnight. Use frost blankets to protect palms that are planted outside during cold winter months.

Humidity. Greenhouses maintain high levels of humidity, much higher than most surrounding areas. Even palms that have been living in a garden center or nursery are surrounded by other plants that maintain a high level of humidity. Spray the foliage of your palms frequently to increase the humidity and make sure the surrounding soil stays moist. Antidessicant sprays will also lock in the moisture in the palm's fronds.

Bringing your palm outdoors

Most palms will burn if immediately planted outdoors. Nurseries typically grow palms in greenhouses or shadehouses which are covered with fabric that filters out up to half of the sunlight. For the first few months after placing your palm outdoors, gradually expose the plant to increasing amounts of sunlight. You can "harden off" your palm using an old sheet or cardboard box. Here are two ways to gradually expose your palm to increasing sunlight:

1. Place your palm outdoors and cover it with a sheet or light fabric. Remove the sheet at the end of each day to allow the palm to receive full sun for an hour. Every other day, increase the amount of time the palm is uncovered by 30 minutes until the palm no longer requires a cover.

2. Alternatively, leave the sheet on all day. Every other day, cut small holes in the sheet to increase the palm's exposure to sun. Cut a hole in the top to allow the newest emerging leaf to receive full sun. Your sheet will look like Swiss cheese when it it ready to be removed.

You can dramatically improve the quality of your soil by adding nutrient-rich organic material.

Benefits of mycorrhizal fungi

Do soil optimizers that include mycorrhizal fungi improve palm growth? Yes, according to recent studies printed in *Palms*, the journal of the International Palm Society. A study of cultivated Canary Island date palms (*Phoenix canariensis*) showed increased plant growth after twelve months of innoculation with the mycorrhizae *Glomus mosseae*.

Mycorrhizal fungi increase the surface area of the roots up to ten times its normal reach, dramatically improving the plant's ability to uptake nutrients. These beneficial fungi are available at garden centers and online retailers.

Turning dirt into soil

Preparing your garden's soil before you plant your palm

Your garden's soil will determine your success with palm trees. As a general rule, palm trees need soil that drains well. While some species of palms, including *Cyrtostachys renda*, are nearly aquatic, the majority of landscape palms will suffer in water-logged soil. Soil can also contribute to the cold hardiness of many species. In subtropical areas, cold, damp soils are more damaging to palms than occasional frosts.

Texture. Soils are usually categorized by their texture, typically sandy, silt or clay. Sandy soil contains large grains of particles and drains well. Silt has smaller particles and drains relatively quickly. Clay soil is composed of tiny particles that bind together. Clay soil holds water and nutrients, but does not allow air to penetrate.

The best soil is *loam*, a mix of sand, silt, clay and organic matter. Organic matter helps soils moderate the level of moisture, nutrients and air a palm's roots need to thrive.

Determining soil texture is easy

Your own hand is the best gauge of soil texture. Pick up a handful and squeeze it. If the soil feels gritty and falls apart, your soil is sandy. If it forms a tight, slippery clump, your soil is predominantly clay. If it is slightly crumply but holds together, your soil is close to loam, the most desirable type of soil. Loam is dense enough to retain moisture and loose enough to allow air to penetrate. Crumbly soils are called *friable* soils.

pH. Soil pH is a measurement of the alkalinity or acidity level in the soil. Most palms prefer a pH level of between 6 and 7, although some palms will thrive in alkaline soils.

Salinity. Typically, soils in areas with low rainfall contain excess accumulation of salt. Salt inhibits a palm's ability to absorb water, resulting in leaf tip burn and eventually decline. Some palms, however, are tolerant of high salt levels, including coastal palms like the coconut. During your dry season, leach out excess soils by thoroughly deep-soaking your palms every month.

Drainage. Palms require soil that drains well to survive. Your palm's roots need air to allow them to absorb nutrients and water from the surrounding soil. Without air, your palm's roots will mold and eventually rot.

Alternatively, if the soil drains too quickly, your palm's roots will not receive the nutrients they need. Without frequent waterings, your palm's roots will dessicate from the lack of moisture in the surrounding soil.

Improving Your Soil

Adding organic matter is the best way to improve the texture, drainage and pH of your soil. As it decomposes, organic matter breaks down and creates humus, a sticky material that improves the fertility and texture of the soil.

Organic matter should be fully composted before you add it to your soil. As it breaks down, raw, green organic matter temporarily competes with your plant for nitrogen. Some manufacturers fortify their organic amendments with nitrogen to compensate for this.

How much amendment? To gain the maximum benefit, add 25-50% organic material to your existing soil.

Manure. Well-rotted manure is a good amendment because it contains high levels of nutrients, but typically contains more salt. Make sure the manure is aged well to avoid adding excess salts.

Peat. Excellent soil additive, but it may lower the pH in your soil. Dry peat is sometimes difficult to moisten. Add a wetting agent or presoak the peat in water before adding to your soil.

Coir. The by-product of the coconut palm, coir is similar to peat in its performance as a soil additive. Coir is made from shredded coconut husks.

Wood products. Wood chips, shavings and sawdust should be bolstered with nitrogen as they are still in the process of decomposing and will rob your palm of much needed nitrogen.

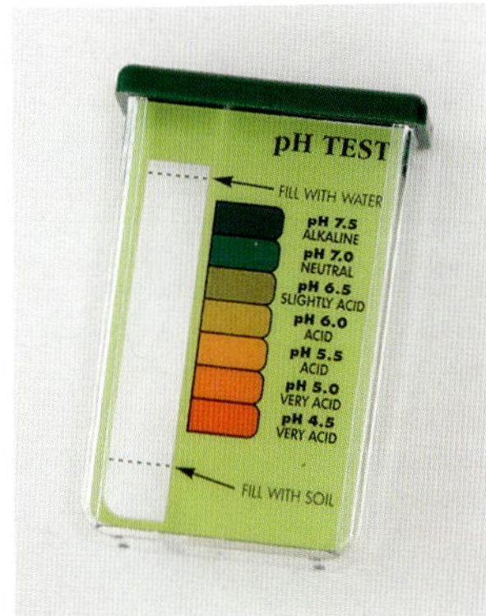

Test your pH

Most garden centers and home improvement centers sell inexpensive kits that will allow you to test the pH in your soil. Most palms prefer a pH level of around 6.5.

Replacing your soil

If your soil is inadequate and can't be improved by amending, consider purchasing soil in bulk from a local landscape supplier. Make sure you purchase soil that has a pH of 6.0 to 6.5. If your water is alkaline, decrease the pH slightly; if your water is slightly acid, increase the pH to compensate. Here is a good mix to request:

- 20% sand
- 10% decomposed granite or cinders
- 40% topsoil
- 30% organic material (fir shavings, fine bark, sawdust or coir)

If your soil mix is too alkaline, add a top dressing that contains organic material that is rich in sulfur. Modify acid soils by adding dolomite lime.

Before planting, thoroughly compact the new soil to remove excess air. Deep waterings will also help your soil settle to the appropriate level. Over time, the organic material in any soil mix will begin to decompose, so consider mounding your beds slightly to compensate for the eventual loss of soil mass.

Planting your palm

A five-gallon size palm is easy for a single person to plant. Large palms may require additional help or equipment.

Materials List:

You only need a few simple gardening tools to plant your palm.

- **Shovel.** Make sure you have a good shovel with a sharp spade tip. You may also need to use a digging bar if you are planting a large palm.
- **Fertilizer.** Buy a good fertilizer especially for palms to place in the planting hole.
- **Epsom salts.** Epsom salts help stimulate root production, providing addtional magnesium.
- **Water.** It seems obvious, but make sure your palm has access to water. For the first several months, make sure the soil stays continually moist, not soggy.

Successful planting means knowing when, where and how to plant

The best time to plant your palm is usually from spring to midsummer, unless you live in a warm winter area. Make sure night temperatures do not dip below 55°F/13°C before you plant your palm in the ground. Roots grow when the soil is warm. If you wait too long, your palm may not have enough time to establish itself before the cold weather sets in.

While some sources still say you should amend your native soil with high levels of organic matter, most palm authorities recommend against adding organic matter to the planting hole, unless you have replaced your soil because of poor drainage or other inhospitable conditions.

Determining the drainage

Dig a two-foot deep hole in the area you plan to plant. Fill the hole with water and wait for it to drain out. Then fill the hole again. If the hole empties in an hour or two, your drainage is good. If the water remains for several hours, your soil drains poorly, and you should amend the soil with organic matter. Worst case: Try another location or create a raised planting bed.

Follow these simple steps to ensure your palm gets off to the right start:

1 Dig a hole that is the exact depth of the container and twice as wide as the opening. Never plant too deep. Palms are sensitive to planting depth; trees planted too deep may suffocate or rot. Trees planted too shallowly will form weak roots that dry out and burn.

2 Place fertilizer in the hole before you position your palm. Adding fertilizer to the hole (rather than the surface) will prevent excess weeds from growing around the newly planted tree. Use caution: Fertilizers can burn sensitive roots, so spread a layer of soil over the fertilizer.

3 Place the palm in the center of the hole. Make sure the base of the palm's trunk is level with the surrounding ground. Use a stick or the handle of a shovel to check. If the hole is too deep, add soil and tamp down firmly to remove any air pockets. If any soil falls off the root ball, place it in the hole. Fill in around the root ball.

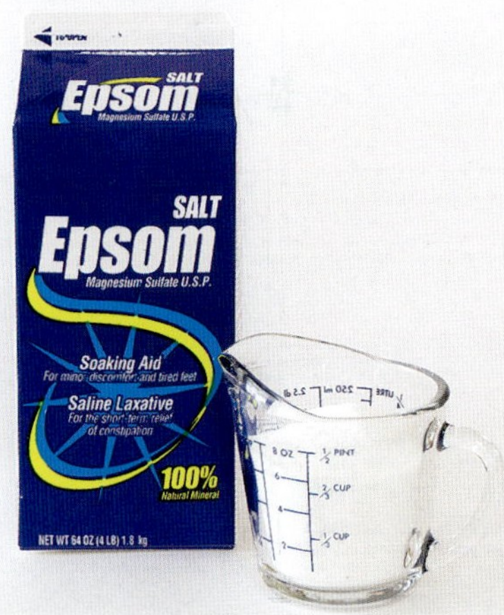

4 Spread a cup of Epsom salts on the soil. Epsom salts are a granular form of magnesium, important in the root production of palms. A cup per palm contains enough magnesium to help your tree establish its roots in its new environment.

5 Water deeply. A thorough watering will settle the soil and remove excess air pockets. Water the palm again in a few days. New and disturbed soils sometimes repel water, so check on the palm between waterings. Keep the soil moist, not soggy, for the first four months.

6 If you live in an area with high winds, brace the palm with 2' x 4' lumber. Strap short lengths around the trunk with metal water heater straps, then nail supporting braces to the strapped wood. Cut the ends of the braces at a 45° angle to make them easy to attach. Never nail directly into the tree; unlike many trees, palms cannot recover from trunk damage.

Choose fresh, ripe fruit

Make sure the fruit you choose is ripe. Depending on the species, the fruit may be red, orange, white, black or blue. Green, unripe fruit can sometimes be coaxed into germinating if you allow it to ripen in a cool, dry area for a few days. Examine the fruit for insects, mold or disease. If you are purchasing seed from a supplier, ask how old the seeds are. Some seeds lose their viability after just a few weeks.

Growing palms from seed

Palms are relatively easy to propagate from seed, but germination (the time it takes for seeds to sprout) may take several months or more. In nature, seed coatings are often impervious to water and gases. Nature breaks down the seed coating, creating breaches in the seeds surface. Called scarification, this process ends the seed's dormancy and allows it to germinate.

Commercial growers sometimes soak seeds in sulfuric acid to accelerate scarification, but most palm seeds will germinate effectively with the techniques shown here.

Fresh seeds are more likely to sprout than older ones. Avoid any seed that looks or smells old or decayed.

The easiest way to sprout palm seeds is the "plastic bag" method, and it's also one of the simplest. Here's all you need to ensure success:

- fresh seed
- plastic gloves
- large, plastic storage bags with plastic zippers, such as Ziploc© brand
- fungicide
- sterile planting medium, such as sphagnum moss, peat moss, perlite or vermiculite
- window screen
- water

1 If your seeds have not already been cleaned, remove the fleshy pulp and fiber from the seeds. Soak the palm fruit in water for 24-72 hours, changing the water daily. Soaking will ferment the fruit and make it easy to remove the fleshy pulp. Keep the seeds in a warm location during the soaking period. Wear gloves when handling the fruit. Many palm fruits contain stinging crystals of calcium oxalate that will irritate your skin.

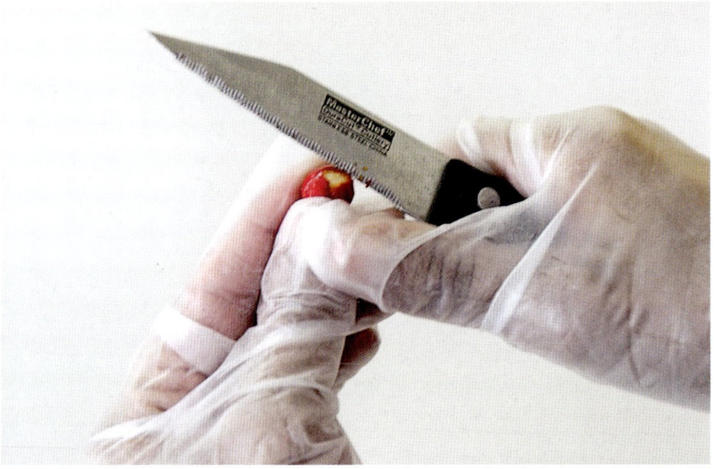

2 Remove the seed from the water and remove the fleshy fruit. Use a sharp knife to carefully peel away the flesh from the seed. To clean a large quantity, use a rock tumbler with gravel and water to abrade the flesh. Wash away any remaining pulp.

3 Soak the cleaned seeds. After you have completely removed the fleshy pulp, soak the seeds again in warm water for another 24 to 72 hours. During this time, most viable seeds will sink. Remove and discard any floating seeds.

4 Soak the planting medium in water to make sure it is completely saturated. Some planting media, such as sphagnum moss, make take several hours to fully absorb water. Then squeeze the planting media with your fists to remove excess water. The planting media should be moist, not wet. If the planting media is too wet, it will likely grow mold, not palms!

5 Place damp, not wet, planting media and palm seeds in the plastic bag and seal tight. Then place the bag in a warm, dark location. If possible, place the bag in an area where it can receive bottom heat, such as on a water heating tank or on an electric heating mat. Most palm seeds germinate quickly in temperatures of about 80°F/30°C.

6 Check for progress every few days. If the media seems to wet, leave the bag open for a few hours. If the media is too dry, mist with a mixture of water and fungicide. Your seeds may begin to sprout in as little as one week, although some seeds may take more than a year.

7 Carefully remove sprouted seedlings from the bag and plant each in a tall narrow container. If the seed is still attached, do not remove it. The seedling receives nourishment from the seed for months. Use a well-draining medium (a mixture of perlite or vermiculite, with peat or sphagnum moss) and keep the soil moist, but not wet. Place your seedlings in bright, indirect light.

How much water?

The amount of water you give your palm depends on a number of factors, including:

- **The age of the tree.** Young trees and new transplants require frequent waterings.
- **Temperature.** The higher the temperature, the more frequent the watering.
- **Humidity.** Increase your watering frequency in dry air climates.
- **Wind.** Wind desiccates the fronds of most palms. Frequent watering can reduce stress.
- **Shade.** Plants in shade typically require less frequent watering than palms planted in full sun.

Two rules to follow: Give palms a deep soak rather than frequent sprinklings; keep the ground moist, not wet, around your palm.

Palms are easy to care for, but they have unique cultural needs. Deep waterings, regular fertilizing and basic maintenance will keep your palm looking great for years.

Maintaining your palm

Palms are rather easy to grow. Like all plants, they have cultural requirements for food and water. They get few pests and, within the group, they can withstand a wide range of climatic and environmental differences.

Observing your palms. The best way to know if you are providing the right care for your palm trees is simple: look at them. Most people wait until their palm is in the final stages of decline before noticing any problems, so examine your palms every

Drip irrigation

Palms prefer long, deep soaks, but many people plant their palms in turfgrass, where they receive frequent, short sprinklings. Drip irrigation systems allow you to control the duration and amount of water your palm receives. These inexpensive systems can easily be installed by the average homeowner and are available at most garden centers and home improvement stores.

month or so. Examine the leaves closely. They should be green and devoid of any stripes and speckles. Look closely for insects. While palms are not particularly susceptible to pests, spider mites and scale may be present. Examine the trunk for signs of distress or decay.

Watering. There is no single formula for watering a palm, but most growers recommend long, thorough soaks rather than frequent light sprays. Many homeowners plant their palms in their lawns and expect their grass irrigation system to water their palms. But most sprinkler systems only penetrate the first few inches of soil. Palms planted in turf grass may require supplemental water. The best way to know if your palm is getting enough water is to dig a small, deep hole about 2 ft./.5 m. away from the palm's trunk. Check to see if the water is penetrating the surface and reaching the root zone.

To provide deep soaks for your palms, most growers recommend installing a drip irrigation system that delivers water directly to the root zone around the trunk. The best alternative is a weekly soak with a garden hose.

Some palms will eventually tap into underground water sources, allowing you to dramatically reduce your watering regimen.

Mulching. Mulching around your palm not only looks nice, it helps your palm in several ways by:

- Helping the soil retain moisture
- Regulating the soil temperature
- Reducing the spread of weeds, which steal water and harbor insects
- Improving the soil composition as it breaks down

Clear away any grass or weeds that are growing within 2 ft./.5 m. of your palm's trunk. Dig out any sod to reduce the chance of regrowth. Spread a 2 in./5 cm. layer of organic mulch around your palm's trunk. Pine bark, wood chips, grass clippings and nut shells make excellent mulches. Make sure your mulch is clean.

Weeding. Weeds will rob the soil of water and harbor insects, including grasshoppers which spread lethal diseases. Hand-weed around your palms or use one of these methods:

1 Weed fabric. Weed fabric will allow water and nutrients to penetrate the surface, but prevent weeds from sprouting. Cover the weed fabric with mulch to improve the appearance. Never lay plastic down around your palms; plastic will reduce the water flow to the roots.

2 Herbicides. Most herbicides can be used without concern around your palms. To prevent overspray, apply during calm wind conditions and cover your palm with a sheet or a piece of cardboard. Use care when applying chemicals and be sure to read the safety instructions on the label.

Clockwise from top: Mulch helps keep turfgrass and weeds away from the trunk of your palm; weed fabric prevents weeds from growing, but allows water and nutrients to penetrate; herbicides can be sprayed close to your palm to kill weeds. Most palms will not be harmed by herbicides, but a piece of cardboard or an old sheet can be used as a shield to deflect any overspray on windy days.

Problem with yellow fronds?

Palms need more potassium and magnesium than most plants. In many areas soils lack the basic nutrients your palm needs to survive. The most common problem with landscape palms is a magnesium or potassium deficiency. Good palm fertilizers have a balance of the nutrients your palm needs to stay green.

Palms require a unique fertilizer mix to ensure they grow and maintain a lush, green appearance. This palm is suffering from a magnesium deficiency.

Fertilizing your palms

A recommended schedule to keep your palms green

Most palms are heavy feeders and require regular fertilizing. Several commercial manufacturers have developed fertilizers that will ensure your success.

The best formula for palms

The formulas for plant fertilizers are represented by a ratio of nutrients. The best palm fertilizers have a ratio of the following:

- 3 parts nitrogen (N)
- 1 part phosphorus (P)
- 3 parts potassium (K)
- 1 part magnesium (Mg)

April	June
August	October

When to fertilize

To maintain a consistent amount of nutrients in the soil, fertilize your palms frequently. Most palm authorities recommend fertilizing when the ground begins to warm in the spring. Start fertilizing your palms in April, then reapply every other month through October. If you live in a warm winter area, like Hawaii, continue throughout the winter.

How to read a fertilizer label

Most complete palm fertilizers contain the three primary macronutrients plants need: nitrogen (N), phosphorus (P) and potassium (K). Fertilizer labels always display these nutrients using the following format, which shows the percentage of the nutrient by weight:

3_N-1_P-3_K

So, 3-1-3 contains three percent nitrogen, one percent phosphorus and three percent potassium. Palm fertilizers should also contain one percent magnesium and include manganese, among other micronutrients. You can boost individual nutrient levels with additional fertilizers. For example, Epsom salts can be added to boost magnesium levels.

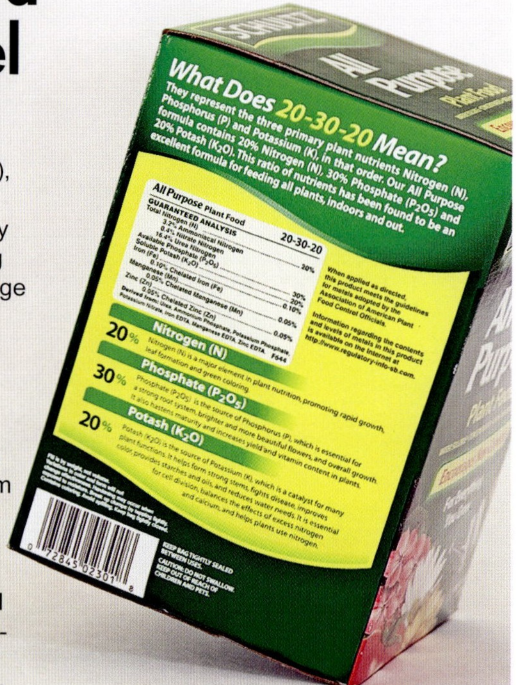

Potassium deficiency. In older leaves, potassium deficiency can be identified by leaves which appear translucent yellow with speckles or spots of dead tissue. Additional potassium sulfate will prevent this problem in new leaves.

So, a 3N-1P-3K-1Mg formula is the best ratio for most palm trees. 18N-6P-18K-4Mg is a common ratio. The best palm fertilizers are called "controlled-release" formulas. Compared to the water-soluble formulas, these fertilizers remain in the soil and release nutrients over an extended period of time.

Fertilize Four Times Each Year

To keep your palms green, fertilize four times per year, beginning in the spring. Here's a good schedule to keep:

- April 1
- June 1
- August 1
- October 1

If you are using controlled-release fertilizers, read the label and modify your schedule based on the instructions. some fertilizers release nutrients over a nine-month period, allowing you to reduce the frequency of application.

Change fertilizers in the fall. During the fall as growth begins to slow, your fertilizer mix should contain limited nitrogen and high concentrations of potassium. In cold-winter areas, many fertilizers may stimulate new growth that is damaged by cold weather. Review the "Winter Care" section for suggestions on how fertilizers will improve your palm's cold-hardiness.

Fertilize the roots, not the trunk. Your palm's roots spread out for many feet beyond the trunk, so you need to apply your fertilizer far beyond the trunk. Rather than dumping a few handfuls of fertilizer around your palm, it's better to broadcast your palm fertilizer in a wide band around your palm trees. Spread 1.5 pounds of palm fertilizer per 100 square feet in the growing zone. Surrounding plants will benefit from the additional fertilizer they receive.

In Hawaii and other tropical areas, spread one pound of fertilizer per 100 square feet every month, particularly if your soils are volcanic or sandy. Again, read the label to determine the recommended frequency of application.

Magnesium deficiency. Commonly called "curly top," this deficiency can be identified by leaves which have broad bands of yellow on the outer margins of the leaves. The tips become brown, but the center margin stays green. Common Epsom salts can be sprinkled frequently around the root zone to increase the magnesium in the soil.

Moving your palm indoors

If you are planning to move your palm indoors for the winter, you can begin to acclimate the plant by bringing it indoors at night. If your palm is in a container, the task is simple. If your palm is growing in the ground, you must transplant the tree into a large container.

Digging up your palm

Prior to digging, make sure you thoroughly wet the soil around your palm to help keep the root ball intact. The root ball should be at least 8 in./20 cm. greater than the trunk and at least 18 in./14 cm deep. Carefully dig the root ball and place it in a container with drainage holes.

Many palms can survive temperatures below freezing. The windmill palm *(Trachycarpus fortunei)* can survive snowstorms and temperatures down to about -6°F/-21°C.

Winter care

Strategies for palm growers in cold climates

Winter is typically a stressful time of the year for palms, even in tropical and subtropical locations. Many palms rest or go dormant for weeks or months during these months.

In cold temperate climates, you can dramatically change the temperature around your palm to ensure it can survive through the winter. Even if your palm loses its fronds, it will survive if the roots and the growth bud (near the top of the palm) are protected.

Remember, site location is the most important factor that will determine your success. Make sure you have planted your palm in the most favorable location on your property. Listed below are several methods to change the microenvironment around your palm.

Small, low-wattage Christmas lights will increase the temperature around your palm by several degrees, particularly when the air is calm. Wrap lights around your palm closely to increase the warming effect. Examine your palm during the day to ensure the bulbs have not burned the foliage.

Water. Make sure your palm gets ample water during the winter. Sun and wind will take away needed moisture from your palm through the cold months. If you live in an area that does not get winter rains, water your palm every two weeks throughout the winter.

Water will also warm the soil and may keep the ground several degrees warmer through the cold night. If you are expecting a frost, make sure to water the ground before the frost. Make sure you do not water the growth bud or fronds of the palm. Water can freeze and damage this area of your palm.

During a freeze, sprinklers can be left on throughout the night. When the temperature freezes, ice will form on the fronds and trunk, thereby insulating the palm from colder temperatures. The temperature inside the layer of ice will not drop below 32°F/0°C. If you are going to employ this technique, first spray the growth bud with a fungicide and cover with a temporary plastic wrap to ensure water does not damage this sensitive area of the palm. Remove the plastic before the sun hits the palm in the morning to prevent the palm from burning.

Snow. Believe it or not, snow is a good insulator for your palm. Snow will protect your plant from sub-zero winds and heat radiation. Monitor the thickness of the snow, as its weight may damage the fronds of your palm. Wipe off any excess snow to prevent injury to the plant.

Mulching. Depending on the severity of your climate, mulch may be the only winter protection your palm needs. Spread a thick 6 in./15 cm. layer of mulch around your palm. Keep the mulch from touching the trunk of the palm to prevent the formation of fungus. The mulch should contain organic material, such as well-rotted manure. As it decomposes, the mulch will generate heat and help the soil retain extra warmth for a short period of time.

Fungicide. Palms grown in cold, wet climates are particularly susceptible to crown rot. When water sits in the ter-

Antidesiccant spray

These chemicals can improve the cold hardiness of your palm by up to 6°F/-14°C. Biodegradable products like Wilt-Pruf form a waxy coating on the surface of the palm and an act like a plastic wrap, protecting against freezing winds and reducing moisture loss. Apply at least two coats, allowing coats to dry between applications.

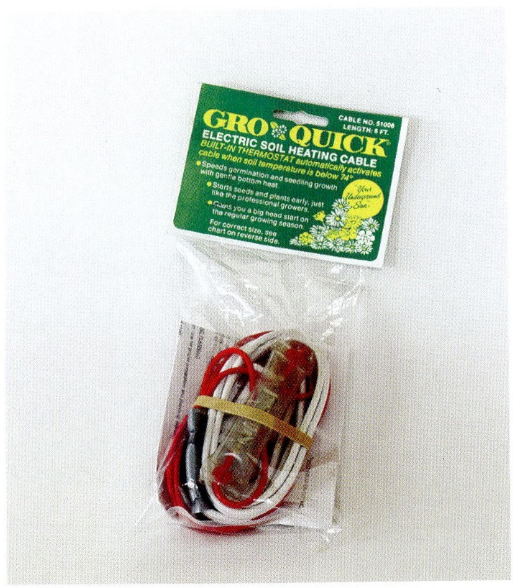

Soil warming cables

Low wattage cables can be wrapped around the root ball of your palm before you plant. This is an effective way to ensure your palm's roots can continue to uptake water during the winter. Purchase these cables at your local garden center or through a mail-order garden supply company. Make sure you read the manufacturer's instructions before working with electrical cables.

minal bud of the palm, fungus and decay set in and will destroy the plant. Fungicide should be sprayed on your palm every two weeks to help prevent damage or death from fungus.

Potassium spray. Liquid forms of potassium silicate will protect and strengthen the plant tissue of your palm. Potassium silicate should be sprayed on the palm at least a day before a frost. Spray in the morning to ensure the plant absorbs the product. Potassium silicate can be applied every few weeks during the winter to further strengthen the plant.

Antidesiccants. Antidesiccants may improve the cold-hardiness of your palm by up to 6°F/-14°C. These biodegradable products form a waxy coating on the surface of the palm an act like a plastic wrap, protecting against freezing winds and reducing moisture loss. Apply at least two coats, allowing coats to dry between applications. These products wash off, so apply every month throughout the winter.

Trisodium phosphate. Trisodium phosphate, or T.S.P., is a cleaning agent that is available in hardware stores or supermarkets. When T.S.P. is added to water, it forms a chemical reaction that releases heat. The effect lasts for about four hours. To use, place several cans of water around your palm. When the temperature reaches freezing, add the T.S.P. to the water.

Umbrellas. Large patio umbrellas will help prevent frost damage on cold nights. As heat rises from the soil, the umbrella will help trap and retain it, keeping the air warmer than the surrounding environment. This technique only works during evenings when the air is calm. Open the umbrella during the evening and close when the sun rises.

Frost blankets. Frost blankets can provide significant protection against cold temperatures, and they also allow light to reach the palm's fronds. Frost blankets can provide more than 8°F/-13°C of freeze protection. Plants must be covered all the way to the ground. Frost blankets should be pinned to the ground with large metal staples, also available at various garden centers.

Cages. One of the best ways to provide winter protection is to construct a wire cage around your palm and fill it with leaves or straw. Use concrete reinforcing mesh to create a closed tube around your palm. The cage should be taller than your palm and wider than the fronds. The cage can be secured to the ground using metal fence posts. Before assembling the cage, spray your palm generously with a fungicide to prevent diseases. When the cage is in place, you can fill it with leaves or straw, which will permit air flow and

Three ways to protect palms

If you want to grow palm trees outside their normal hardiness zone, you can dramatically increase your palm's ability to survive cold climates by providing winter protection. In some cases, temperatures inside these structures may remain up to 20 degrees higher than the surrounding areas. These strategies work well for palms that do not exceed 6 ft./2 m. in height.

Insulate with straw. Build a simple wire cage around your palm with wire fence material and wooden or metal stakes. Spray your palm with fungicide, then fill the structure loosely with straw or leaves.

Add a frost blanket. To add more winter protection, wrap the structure with a frost blanket, available at nursery supply stores. Frost blankets add several additional degrees of frost protection.

Wrap with plastic. Plastic wrap will add at least 10 degrees of cold protection to your structure. Plastic retains moisture, which can promote fungal growth. During warmer days, open the wrap to ventilate.

Frost blankets provide significant protection from cold temperatures, adding several degrees of freeze protection. As an added benefit, these special fabrics admit daylight, allowing you to leave the blanket on the palm for the entire season.

provide filtered light. As the organic material decomposes, it will produce added heat. Your cage can also be lined with a frost blanket or plastic to provide added protection.

Artificial light. Wrap Christmas lights around your palm to increase the surrounding temperature by several degrees. Wrap the lights close together to increase the effect, but make sure the lights are not touching the foliage of the plant or any flammable material. Larger incandescent bulbs can also be used to increase the temperature around the palm. Make sure lights are plugged into an outlet with a ground-fault circuit.

Heating cables. Low-wattage soil warming cables, available from garden centers, can be wrapped around the root ball of your palm before you plant it. The cables can be turned on during cold spells to warm the ground temperature above freezing. Make sure you read the manufacturer's warning before using these cables.

Cardboard boxes. A cardboard box can act as a temporary coldframe. Cover your small palm when the temperature is expected to dip below the minimum survival temperature for your palm. Cardboard shelters are ideal for areas that do not experience prolonged cold periods.

Miniature greenhouses and coldframes. You can easily create a winter greenhouse for your palm with wood and plastic. Use 2x4 pieces of lumber to construct a wood frame around your palm. Build a frame that is taller than your palm and wider than its fronds. Use screws to fasten the wood, then disassemble the structure when the weather improves. Fasten a double layer of plastic to the wood frame using a staple gun or small nails. Create a door on one side that can be opened to allow ventilation as needed, particularly on sunny days. Leave excess plastic at the bottom of the frame to create a tight seal. The excess plastic can be covered with soil or mulch. In areas of heavy snow, slant the roof and cover with rigid plastic or plywood.

Unheated greenhouses are not effective on windy or extremely cold nights. To further protect your palm, cover it with a frost blanket before assembling the greenhouse. Place incandescent lights inside the structure to add heat. Make sure the lights do not touch the palm's fronds or any flammable surfaces. Wrap Christmas lights around the palm to further increase the temperature.

After a freeze

If you live in a frost-prone area, you should take special precautions with your palms. After the last freeze of the winter, you can begin to prepare your palm for the growing season. Cut off any fronds that are damaged, and pull on the fronds at the apex (growth bud) of the tree to check the viability of the plant. As a precaution, spray your entire palm with fungicide at least once after the final freeze of the year.

Make sure your palm gets ample water during the spring, but be careful not to overwater your tree. As a rule, palm trees prefer moist, not soggy, soil.

If it's brown, cut it down

Only remove dead or diseased fronds from your palm. Most people make the mistake of trimming away viable fronds from a palm's canopy. As their fronds begin to yellow and die, palms translocate nutrients to other parts of the plant. Cutting these fronds before they have turned brown will rob the tree of valuable nutrients and slow the palm's growth.

A telescopic pruner with a saw blade makes it possible to remove dead fronds from this queen palm *(Syagrus romanzoffiana).*

Pruning

Palms with crownshafts, such as royal palms (*Roystonea*) are considered "self-pruning." As their fronds die, they simply fall to the ground. Others, like the Mexican fan palm (*Washingtonia robusta*) retain dead fronds and form a "petticoat" of brown leaves.

Pruning palms is often a matter of aesthetics. Depending on your preference, you may prefer to remove the dead fronds from your tree or leave them alone for a more natural appearance. Hygiene is another consideration: Rodents and insects often nest in the dead fronds of palms.

In general, you should only remove brown fronds from your palm trees. Palms translocate nutrients from dying fronds to new growth, so removing these fronds prematurely will deprive the tree of these vital nutrients.

Avoid the feather duster look

The Canary Island date palm in this photo has been severely pruned by a tree maintenance crew. Many professional tree trimmers overtrim palms to save money by lengthening the duration between trimmings. This tree would look more attractive and grow more productively with twice the number of fronds. Trim only dead, damaged or diseased fronds.

Pruning Rules

To promote growth and vitality in your palms, follow these basic rules:

1 Only remove brown fronds. Your palm reabsorbs nutrients as older fronds fade.

2 Prune during the summer months. Wounds late in the season may promote fungal growth.

3 Use sharp pruning shears. Carefully cut the fronds close to the trunk. Some trees, such as those in the *Phoenix* family, may require the use of a large pruning saw or professional care.

4 Disinfect pruning equipment between trees with a 10-minute soak in a 5% bleach solution. Dirty equipment spreads lethal diseases.

5 Never remove fronds that are held above a horizontal plane. Only remove fronds that hang down or are at least 50% brown.

Trim leaves that drop below an imaginary horizontal line. Trimming branches that remain upright destroys the aesthetic appeal of the palm.

Tools of the trade

Three simple tools will make it easy to maintain the appearance and vitality of your palms. Keep you tools sharp and well-oiled to ensure success.

Bypass pruners. These pruners are ideal for trimming dead fronds. Buy pruners that are large enough to prune thick branches. Keep these pruners sharpened at all times. Ragged cuts made with dull blades may lead to disease and infection.

Linoleum knife. A simple linoleum knife, available at most hardware stores, is a great tool to remove the leaf bases from trees like many fan palms. Use these knives to horizontally score the leaf base.

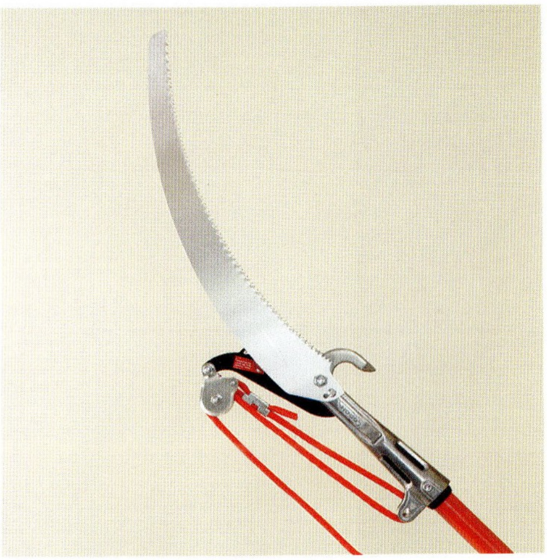

Telescopic pruner. These pruners are essential for removing dead fronds that are hard to reach. Many telescopic pruners will extend your reach by 20 ft./6 m. or more. Buy a telescopic pruner with a rigid saw blade; some pruners have flimsy blades.

Trimming a Canary Island date palm

Shaping a Canary Island date palm requires careful planning and an artistic eye. Professional tree trimmers use chainsaws to prune palms with thick fronds. Chain saws are hard to disinfect, however, and may spread disease between trees.

1 Using a sharp saw, fronds are removed. The fronds are carefully cut to the top of the trunk. The saw blade should be parallel to the ground; do not angle the saw blade toward the trunk. The leaf bases should have a uniform appearance when finished.

2 About 3 ft./1 m. from the crown, mark the trunk to delineate the size of the final "pineapple." The growth bud of the tree is located near the top of the tree and cannot be damaged or the tree may decline or die.

3 Skin the trunk of the tree by cutting several inches deeper toward the trunk. At this point, you will establish the ultimate diameter of the trunk.

4 Begin shaping the pineapple by cutting the remaining leaf bases at a 45° angle.

5 If possible, step back and examine the tree from a distance to ensure a symmetrical cut.

6 The final tree will look similar to the palm in this photo. Carefully clean any remaining plant material to finish the project.

The many looks of the Mexican fan palm

The Mexican fan palm (*Washingtonia robusta*) is famous as the symbol of Los Angeles' tree-lined streets. In California, most municipalities trim away the fronds and skin the trees of the remaining leaf bases, or "boots."

In its natural habitat, the Mexican fan palm, like many other fan palms, retains its dead fronds, which are sometimes called a "petticoat." In cultivation, however, these dead fronds harbor insects and rodents and can create a fire hazard.

Clockwise from top left: Trimmed leaf bases create a lattice effect, a compromise between a trimmed trunk and a natural appearance; trunks skinned of leaf bases; dead fronds form a "petticoat" that covers the trunk.

Trimming a Washingtonia

Tools Needed:

- Pruning shears or saw
- Linoleum knife

Depending on your preference, you can prune many fan palms to create a smooth-trunked palm or a lattice effect. The latter is created when you retain the closely trimmed leaf bases, or boots, of many fan palms, including cabbage palms (*Sabal palmetto*) and Mexican fan palms (*Washingtonia robusta*). Skinning involves adding an additional step to the pruning process to remove the leaf bases.

1 Remove the dead fronds from the tree using a saw or pruning shears. Trim as close to the trunk as possible.

2 Starting at the bottom of the tree, slice vertically through the leaf bases with a linoleum knife.

Regenerative roots

Unlike many other trees, palms can regenerate their roots. Depending on the species, severed root stubs will branch and form new roots or drop off completely. Mature *Sabal* palms are relatively easy to transplant because the roots and the remaining soil can be almost completely removed. Within months, the tree will regenerate roots directly from the trunk.

Many palms, like Bismarck palms *(Bismarckia nobilis)* benefit from root pruning. About 60 days before transplanting, dig a sharp spade into the ground about 12 in./30 cm. away from the trunk. The palm will begin regenerating new roots which can uptake water after final digging.

The fronds of these cabbage palms (*Sabal palmetto*) have been removed to help preserve the tree's stored water. Certain genera, like mature *Sabal* palms, can withstand severe transplation shock if the fronds are entirely removed.

Transplanting

Palms are among the easiest large plants to transplant if you follow these simple rules

Most mature palms are easy to transplant. Unlike many other trees, you can transplant a sizeable tree with a relatively small root ball. Here are a few pointers to ensure your success:

Time of year. Transplant at the beginning of the warm season. The soil is beginning to warm, stimulating root growth.

Root regeneration. Palms regenerate new roots to replace those that have been cut. Typically, a large percentage of the severed roots wither away and are replaced by new roots. Some genera, including *Sabal*, rely entirely on newly regenerated roots, so it is unnecessary to dig and transplant a large root ball.

Digging the root ball. Wet the area around the base of the palm to help the soil cling to the roots. Dig a root ball about 12 in./30 cm. from the base of the trunk. The depth of the hole should be about 24 in./61 cm to 36 in./91 cm. Some palms, like *Bismarckia nobilis,* may require a deeper hole.

Moving the palm. If the soil drops from the roots, wrap the root mass with burlap. Some species of palms with crownshafts, including *Archontophoenix*, have sensitive growth buds. To protect the root bud, splint the crown with 2x4 lumber tied around each trunk. The splint should extend past the top fronds.

Removing the fronds. Unlike most species, *Sabal* palms benefit from having all

For most species, the root ball you dig can be relatively small. Most professional tree growers leave 12-18 in./30-46 cm. inches of roots attached to the trunk. Use a shovel to measure the distance or mark the ground with spray paint to determine the root ball size. Root depth is also important. Dig down about 24-30 in./61-76 cm. to retain as much depth as possible.

The root balls of most palm trees are typically very small in proportion to the height of the tree. To protect the roots and retain a small amount of existing soil, wrap the root ball with plastic or burlap. The remaining soil and wrap will help keep the palm's roots moist. If your palm will be out of the ground for an extended period, water the roots to keep them moist. To help retain moisture and moderate the temperature in the root zone, bank soil around the root ball.

of their leaves removed. Until they regenerate new roots, these palms have no way to uptake water. Leaf removal helps them conserve water until they establish new roots, typically in about six months.

Planting depth. Like palms grown in containers, plant your palm at the same depth that it was previously growing. Never mound soil around the palm's smooth trunk area. Planting a palm too deeply, even by a few inches, can cause decline and death. Similarly, palms planted in a shallow hole will lack the stability they need to grow. New, emerging roots will not reach the soil. To remedy, mound up the soil around the roots.

Soil. In general, transplanted palms should be planted in well-draining soil. Some professional tree companies recommend backfilling the hole around the rootball with pure sand, claiming the newly severed roots are susceptible to disease and decay if exposed to contaminated organic material.

After care. Regular watering is the most critical element to ensure your success. Begin fertilizing after four months when roots begin to grow. New root growth is often accompanied by new leaf growth. Expect your palm to begin producing new fronds within six to 12 months.

Preventing infection

Here are three easy ways to help prevent soil-borne diseases:

1 Make sure your palm's soil is well-drained. Replace the soil in container palms every two years with fresh soil that allows air to penetrate the roots.

2 Don't overwater container specimens. Palm roots require air as well as water to grow. Never let a container palm sit in water, unless it is one of the few aquatic species.

3 Avoid watering your palms from above. Don't allow water to pool around the spear of the palm, particularly young seedlings in containers.

Fusarium **wilt is a problem in many areas, including the United States, France and Australia. The disease is fatal and is usually spread by unsanitary pruning equipment.**

Pests and Diseases

Palm trees are not very susceptible to pests and diseases. Most can be controlled with treatments that can be purchased at a local garden center or home improvement store. In extreme cases, plants must be removed and disposed of to prevent the spread of infection.

Insects. In most cases, insect infestations are controlled by predators, including beneficial insects and small animals and birds. The most common insects, particularly on container palms grown indoors, include spider mites, aphids and scale. In some areas, royal palm bug and other beetles, infest the new growth on palms.

Most aphids and scale can be controlled with soapy water. Spray the entire palm, including the undersides of the fronds. Treat severe insect infestations with systemic pesticides that can be applied as a drench or granular application in the root zone of the palm. Within a few weeks, your palm will translocate the pesticide through its system and make the palm virtually immune to most insects. A drench of the chemical imidacloprid is one of the best systemics to control royal palm bug.

Prevention is the best defense against insects. Make sure your palms are well watered, fertilized and located in a weed-free area. Remove brown, dead fronds that typically harbor colonies of insects. Hose the foliage of your palms with water on a regular basis to remove dust, dirt and insects.

Fusarium wilt. *Fusarium* wilt is a problem in some palms, including the Canary Island date palm (*Phoenix canariensis*). Leaves will decline and wilt, leaving just a few fronds standing erect.

Fusarium wilt has damaged many old avenue plantings in California, France, and Australia. The disease is spread by infested seeds, soil and pruning equipment, particularly chain saws. No cure exists; palms with *Fusarium* wilt must be destroyed. To prevent this disease from spreading, make sure you sterilize your pruning equipment with a 10-minute soak in a 5% bleach solution. Isopropyl and denatured alcohol can also be used to sterilize equipment.

Lethal yellowing. Lethal yellowing is a major problem throughout the world. The disease attacks many genera of palms, including coconut palms (*Cocos nucifera*) and Christmas palms (*Adonidia*). Spread by a small leafhopper, this causes the decline and yellowing of the fronds. Eventually, the spear of the palm will collapse. Injections with oxytetracycline can help prevent or treat the disease.

Ganoderma butt rot. This disease is fatal and usually attacks older palms. Ganoderma butt rot usually establishes itself in a trunk wound and then attacks the entire tree. This fungal disease looks like a large conch shell at the base of the trunk. Remove affected trees and fumigate the surrounding soil. This disease may be spread by mulch that is infected with its spores.

Phytophthora. Primarily rots the roots of palms that are planted in soggy soil, but the disease can also affect the crown. This disease is most common in palms after a damp, cold winter or during the hurricane season. When it attacks the palm's crown, the spear becomes blackish-brown and can be pulled out easily. If the disease affects the roots, the roots will decay and the palm will wilt. The best treatment is an antifungal drench, like copper fungicide, to the crown and roots of the palm.

Left: Aphids are one of the most common insects that attack palms in the landscape. Chemical treatments, including systemic insecticides, will help control insects.

If you prefer an organic approach, use insecticidal soap or natural predators, such as lady bugs.

Scale insects appear as hard brown or whitish spots on leaf fronds. This palm leaf has been colonized by scale, which will eventually kill the frond and the entire tree. Spray scale-infected trees with insecticides or use a systemic control that will penetrate the fronds and make them toxic to sucking insects. Scale can also be removed with insecticidal soap applied with a toothbrush to the infected areas.

Organic alternatives

Synthetic fertilizers can provide the full spectrum of nutrients your palm needs to survive, but they do little to improve the soil in your garden. Rather, they add salts and can sometimes injure beneficial organisms. Organic fertilizers, including cottonseed meal and alfalfa meal, will actually improve the soil and encourage the reproduction of beneficial microorganisms. Organics must be applied more often, but their long-term benefits outweigh those of synthetic chemicals

Palm care seasonal checklist

Easy steps to ensure your palm looks great all year

From simple tasks like trimming dead leaves with scissors or pruning shears to fertilizing, each season brings a small set of tasks to ensure your success with palm trees. Follow this simple season-by-season checklist to show you how to maintain your palms and help them thrive.

Spring

As the ground begins to heat up and the air begins to warm, your palm will begin to grow. As the soil warms, so does your palm's ability to absorb nutrients. Now is the time to perform post-winter maintenance and begin fertilizing your palm.

If your palm has been growing indoors, you can begin to acclimate your palm by gradually exposing it to its new environment. Early summer is typically the best time to plant outdoors, but you may be able to plant your tree outdoors if temperatures remain 55°F/13°C.

☐ **Prune.** Pruning off any dead plant material will improve the appearance of your palm and eliminate hiding spots for pests and diseases. Prune any dead fronds and old leaves that may have been damaged during the winter months.

☐ **Fertilize every 60 days.** Now is the time to begin fertilizing your palm with a granular fertilizer designed specifically for palms.

☐ **Stimulate roots.** Apply a cup of Epsom salts to the soil around your palm. Epsom salts contain magnesium that will accelerate the growth of your tree.

☐ **Spray for insects.** If you see any insects on your palm, spray your palm thoroughly, making sure to coat the undersides of the palm fronds.

Summer

Summer is your palm's favorite season, and it's probably the most important season to ensure your palm thrives in the cooler months. Now is the time to establish a regular routine for watering, pruning, feeding and pest control.

☐ **Water properly.** Most palms require ample water—at least an inch of water a week. Deep soaks are better than infrequent sprinklings. Consider installing a soaker hose around your palm to deep water the roots of your palm. Soaker hoses and drip irrigation are the most efficient method for watering palms.

☐ **Wash foliage.** Supplement your deep soaking with overhead sprinkling. Using a garden hose with a nozzle, apply a strong jet of water to clean the fronds and dislodge any insects. Water in the morning or early afternoon to ensure the palm is dry before nightfall.

☐ **Prune.** Prune your palm to remove any dead fronds. Use pruning shears to remove fronds close to the trunk. Make sure you do not damage the growth bud of the tree. Use scissors to cut away dry or damaged frond tips. Disinfect pruning tools in a mixture of equal parts bleach and water between palms to prevent the spread of disease.

☐ **Fertilize.** Continue fertilizing your palm every 60 days with a fertilizer designed specifically for palms. Make sure you follow the instructions on the label and do not over fertilize. Fertilizers contain salts that can damage and burn your palm. Stop fertilizing with palm fertilizer in August and switch to a fertilizer that does not contain nitrogen. Nitrogen encourages new tender new growth that will be damaged as the cold weather sets in.

☐ **Control pests.** Palms are not particularly susceptible to severe insect damage, particularly if they are healthy. Make sure your palm is not stressed by lack of water or damaging winds. Spray your palms regularly with a strong jet of water to dislodge any insects. Most outdoor insecticides formulated for garden use can be used on your palm to eradicate any damaging insects.

Fall

In early fall, you should begin to prepare your palm tree for the cold winter months. Depending on the severity of your climate, you may consider moving your palm indoors or preparing your palm to survive winter outdoors. Most juvenile palms will not survive freezing temperatures until they have developed substantial root growth.

☐ **Fertilize with potassium only.** Beginning in early fall, switch to a potassium-only fertilizer. When water freezes in the cells of a plant, it expands and bursts the surrounding tissue. Potassium strengthens the cell walls of your palm and displaces water, making your palm significantly hardier in cold weather. Use a granular fertilizer, like 0-0-60, at least 45 days before your first anticipated frost.

Liquid potassium can be sprayed every few weeks during the fall to further strengthen the plant.

Note: Avoid any fertilizer that contains nitrogen. Nitrogen encourages tender new growth which will be damaged by cold temperatures. If your palm is planted in an area surrounded by turf grass, do not apply lawn fertilizer anywhere near the root zone of your palm.

☐ **Mulch.** Apply a thick layer of mulch around your palm to help the soil retain warmth. Spread a thick 6 in./15 cm. layer of mulch around your palm. Keep the mulch from touching the trunk of the palm to prevent the formation of fungus.

☐ **Build a greenhouse or cold frame.** If your winter temperatures are severe, now is the time to start building a structure to house your palm for the winter months. Refer to the section on winter care for detailed instructions.

Increase your palm's cold tolerance

Potassium strengthens the cell walls of palms, making them more resistant to cold weather and freezes. If you live in an area that is colder than your palm's native environment, begin fertilizing with potassium in the fall, continuing through winter. Liquid potassium sprays can be applied every few weeks through the cold months.

Palms in containers

Growing palms indoors

A prescription for disaster

Never allow your palms to stand in water. Most palms will suffer and eventually die from oversaturated soil. If your palm sits in a saucer, add gravel to keep the palm's container from becoming water-logged.

Most palm trees make ideal container specimens. Placing a palm on a lanai or in a sunny room will transform your entire home. And, because of their limited root systems, they require very little floor space.

Whether you are planning to keep your palms outdoors or transfer them inside during colder months, chances are there is a wide selection of species that will suit your desires. But one thing you should remember: There is no such thing as an indoor palm. Palms, like any other plant, are

Palms in the shower?

Salt buildup is a major problem with container-grown palms. Most city water contains high levels of dissolved salts which, over time, build up in the soil. The most noticeable signs of salt buildup are leaf tip browning and a white crust on the soil surface. To eliminate this problem, give your palm a thorough soak at least once a month to leach out the excess salt. Place your palm in a shower and allow water to thoroughly drench the soil. Alternatively, place a large basin under the palm's container and perform a similar soak. If the weather is warm, take your palm outdoors and allow rain or a long soak with a hose to cleanse the soil.

best suited for their native habitat, which does not usually include carpeting or air conditioning. So, the first rule in growing palms in containers is to try to replicate the same conditions the plant would receive in the wild.

Providing the right environment. When you grow a palm tree in a container, you have considerable control over the environment and the elements that are necessary for growth, namely light, humidity, temperature, water, soil and nutrients.

Light. Light is the most important aspect in container gardening. Try to place your palm tree as close to a natural light source as possible. The brightest place in your house is usually near a window, so you should place your palm near a window or door, not in a dark corner. If your palm's leaves begin to droop or develop discoloration, move the plant until you find a better location.

If your natural light is insufficient, supplement with artificial lighting. Incandescent bulbs can burn foliage, so make sure you keep them at least two feet away from the palm's leaves. Fluorescent bulbs will stimulate foliage growth and can be placed within a few inches of the palm's leaves. Fluorescent lighting promotes better foliage color than incandescent lighting. "Grow lights" provide full-spectrum lighting and create an optimal environment for your palms, but are considerably more expensive. Spotlights can also be used to create dramatic shadows around your palms. Spotlights and uplights can also provide additional, localized heat for your palm, particularly if your palm is placed in an unheated porch or sunroom. Palms will benefit with up to 18 hours of lighting a day. The more light your palm receives, the faster it will grow. An inexpensive timer will help regulate the amount of lighting you provide.

Humidity. If you are keeping your palm inside a home or an apartment, the humidity level is probably inadequate. Most palms prefer high humidity, but our heating and air-conditioning systems create an environment with dry air. The kitchen and the bathroom are typically the most humid areas of the house, but they are not usually

Top: Dig your finger into the soil to check if your palm needs water. If the soil is dry beneath the surface, water deeply until water begins to drain out of the container.

Left: Purchase an inexpensive water probe at your local garden center. Dig the probe into the soil every week or so to check the soil moisture. Only water when the soil is nearly dry.

Three easy indoor palms

While there are many palms that can be grown successfully indoors, three species have proven themselves as nearly indestructible. These palms make great houseplants you can give as gifts to novice or first-time palm owners.

Kentia palm (*Howea forsteriana*). Images of Victorian England usually include kentia palms as decorations in dining areas and conservatories. This is a single-trunked palm, but most plants offered for sale are actually clumps of three or four plants. Outdoors, this palm can grow to more than 40 ft./12 m. tall, but indoors it will rarely reach 9 ft./3 m. An elegant, tropical-looking palm that will tolerate bright, indirect light and neglect.

Lady palm (*Rhapis excelsa*). This is a clumping palm with deeply divided palmate fronds, giving them the appearance of fingers. Fronds are deep green and the trunks are covered with a brown fiber. An excellent palm for indoors. These palms will grow to about 6 ft./2 m. tall. Rare variegated forms can be purchased from collectors.

Neanthe bella (*Chamaedorea elegans*). One of the most commonly grown houseplants, neanthe bella (also called a parlor palm) can often be purchased at your local grocery store. Usually seen as small seedlings in 4 in./10 cm. pots, neanthe bella can be grown in a small dish garden or as a container specimen. Can survive in dark, indoor conditions for an extended time, but prefer bright, indirect light. Needs ample humidity to look good. Will grow to about 3 ft./1 m. tall indoors.

Picking a good container for your palm

The most important consideration when choosing a container is drainage. Make sure you plant your palm in a container that has a drainer hole on the bottom, otherwise your palm will die from overwatering.

Consider potting your palm in a terra cotta container. Terra cotta will absorb some of the moisture from the soil and reduce the risk of overwatering. You can then test the dryness of the plant by tapping on the container. If the pot rings, the soil is dry and needs water; if you hear a dull thud, the soil (and the container) are damp.

Most interiorscape professionals use self-watering containers that allow the palm to absorb water from a reservoir. These pots are more expensive, but they will reduce your maintenance time and will continue to water your palm when you are away from home on a vacation or business trip.

Palms prefer deep pots to stimulate root growth. Look for a pot that is proportionally taller than its width. When repotting, choose a pot that is one size larger than the present size; it's easy to overwater a small palm in a container with a high volume of soil.

the most desirable locations for your palm tree. To improve the humidity around your palm, place the palm's container on a saucer or tray filled with pebbles. Keep the pebbles covered with water, but make sure the palm's roots are not standing in water. Never let your palm stand in a saucer filled with water. Your palm will suffocate if it does not have adequate air to its root system.

Also consider grouping other plants around your palm. Together they will create a microenvironment with higher humidity. If you notice brown tips on your palm's fronds, increase the humidity by misting your palm several times a week.

Temperature. Fortunately, the temperature most people find comfortable is adequate for most palm trees. In their native habitat, most palms are accustomed to warm days and cooler nights. If you turn your thermostat down at night, try to maintain a night temperature above 60°F/16°C.

Water. Water your palms once a week, no more. More palms die from overwatering than underwatering. Two important rules for success:

- Never let the soil dry out completely.
- Never let your palm stand in a saucer of water.

Check your palms every week or so to see if the soil is moist. Dig your finger into the soil and test the soil an inch below the surface. You can purchase an inexpensive water probe at your garden center that measures the water content in the soil. If the soil is dry, water the palm deeply until water begins to drain into the saucer. Regular deep soaks are better than daily dribbles of water. In the winter as the days shorten, growth slows down and palms need even less water.

Time to repot?

Repot your palm when the roots begin to fill up the container. The rootballs should pull out of the container in one complete mass, with most of the soil clinging to the roots. As a rule, you should repot your indoor palms about every two years.

Fertilizer. Most indoor palms can survive with a surprisingly small amount of fertilizer. In fact, excess fertilizer will increase the salt buildup in your soil and create a potentially lethal environment. Fertilizer provides the nutrients a palm needs to produce its roots, trunk and fronds. Organic fertilizers are derived from plant and animal material, such as manure, fish emulsion and cottonseed meal. Inorganic fertilizers are derived from chemicals and can be formulated to release nutrients over an extended time period. A good palm fertilizer will contain an adequate amount of nitrogen (N), phosphorus (P) and potassium (K), usually seen as a ratio of numbers. Palms generally prefer a ratio of 3:1:3 with one part magnesium and trace elements of manganese, iron, copper, zinc and boron. When you fertilize, always obey these three rules:

- Always follow the manufacturer's instructions
- Water thoroughly before you fertilize
- Don't overfertilize

Soil. Good drainage is important. Although there is no such thing as a universal potting mix, your soil should be sterile and free of insect eggs, weed seeds and soil-borne patho-

Clockwise from far left: Pumice (or cinders) provides excellent drainage and is used in most Hawaiian mixes; peat is considered one of the best organic soil amendments; sand improves drainage and increases the life of a soil mix; redwood compost provides a coarse, organic medium that allows air to reach your palm's roots.

gens. A good mix should be light enough to provide drainage, but heavy enough to support the tree so it does not tip over. A mixture should contain inorganic material and organic material. Organic material, such as peat and fir bark, breaks down quickly and reduces the vital air space around your palm's roots. Inorganic material, such as pumice, cinders and sand, never break down, ensuring your palm's roots will be able to absorb nutrients.

Most commercial potting mixes contain an adequate mixture of inorganic and organic material. If you are looking for a good readily-available potting soil, cactus mixes make good well-draining soil for container-grown palms. If you are mixing your own soil, use 1/3 sand, 1/3 coarse peat moss and 1/3 perlite or cinders.

No-soil potting mix for growing container palms

If you need a large volume of potting mix, the following mix will be ideal for most palms grown in containers. An ideal mix provides at least 10-15% air penetration to allow your palms roots to absorb nutrients. Organic materials decompose and will become waterlogged after a few years, so any mix must contain ample inorganic material to maintain the soil's porosity. Make sure you plan to repot your palms at least every two years. These amounts will create about a cubic yard (9 cubic feet) of potting soil.

- 1/3 yard (3 cubic feet) of washed, coarse sand, pumice, cinders or perlite
- 2/3 yard of organic material, such as coarse peat moss, sawdust or coir
- 2 pounds of dolomite
- 1 pound of slow-release fertilizer

If your soil mix repels water after planting, mix a wetting agent or a small amount of dish detergent with a gallon of water to penetrate the mix.

Hawaiian Mix

The following mix is widely used in the nursery trade in Hawaii. This mix is ideal if you have access to volcanic cinders.

- 1/3 yard of organic material, such as coarse peat moss, sawdust or coir
- 2/3 yard of volcanic cinders
- 2 pounds of dolomite
- 1 pound of slow-release fertilizer

This private conservatory houses a wide variety of exotic palms, allowing the owner to grow tropical species from around the world.

Sunrooms, greenhouses and conservatories

Since the 19th century, palm collectors have built conservatories to house their collections of rare palms. Today, sales of hobby greenhouses and sunrooms have soared because of a renewed interest in tropical gardening.

The main difference between a greenhouse and a sunroom or conservatory is function. Greenhouses are designed for propagating plants and are more functional. Sunrooms and conservatories are designed for people and the display of indoor plants.

Sunrooms and Conservatories

These "living" rooms are ideal for growing palm trees indoors, allowing you to create growing conditions that simulate your palm's native environment. Your sunroom must be able to provide basic needs for plants and people in all seasons.

Flooring. The flooring in your sunroom must withstand dampness and occasional drenches from overwatering. Concrete, brick, tile and stone make good choices for flooring. Make sure you include drainage in your design.

Ventilation. Your sunroom should include windows or skylights that open. Ceiling fans will improve heat exchange and air circulation to prevent fungal diseases.

Climate control. Typically, sunrooms are extensions of existing living spaces and use the same heating and ventilation systems. But, sunrooms tend to warm and cool faster than other rooms because of the heat loss and gain from their large windows. Consider adding some form of climate control to regulate the temperature in the sunroom.

Water. Extend the plumbing into your sunroom to make it easy to water your plants and increase humidity.

Sun shades. Draperies and solar shade fabrics will help reduce the heat gains and losses and prevent sun scorching. Some sunrooms include solar-activated shades that open and close depending on the light intensity.

Temporary shelter

Even a cardboard box will suffice as a greenhouse to protect small palms in cold weather. When bad weather is forecast, place a box over the plant that is slightly larger than the widest and tallest fronds. Remove the box in the morning to allow sun to reach the plant.

This greenhouse is kept clean and free from weeds that may harbor insects and diseases. Sun shades can be added to reduce light levels for sensitive young palms.

Greenhouses

Whether you are considering a simple hobby greenhouse or a commercial-sized structure, advances in building materials have created a broad range of choices at surprisingly reasonable prices. Simple lean-to structures can be assembled in a weekend by a novice homeowner and will provide a significant amount of space to propagate palms or overwinter specimen-sized plants.

Regardless of size, the basics for growing palms in a greenhouse are the same. Greenhouses allow you to nearly replicate the climate and conditions of your palms in their native habitat.

But, remember: Not all palm species enjoy a similar climate. Palms from the hot, dry deserts of Baja California will suffer in a greenhouse that emulates a humid, tropical rainforest. The blue hesper palm *(Brahea armata)* prefers dry, sunny conditions and can tolerate cold evenings. The humidity-loving sealing wax palm *(Cyrtostachys renda)* will die if temperatures drop below 50°F/10°C. Depending on the size of your greenhouse, you may be able to group or partition your palms to suit their individual requirements.

Greenhouse and large stone pillars inside of renowned St. Louis Botanical Garden.

Greenhouse checklist

Ideas to help you plan and maintain a winning hothouse

Greenhouses offer an exciting way to grow palms outside of their normal climate zones, allowing you to replicate nature. Whether you are building a commercial structure or a simple lean-to structure, this checklist includes the essential elements for success:

☐ **Temperature.** A good greenhouse design includes a heating and cooling system and a thermostat to regulate the temperature. Gas furnaces help regulate winter temperatures. Temperatures in your greenhouse should not drop below 55°F/13°C; depending on your region, heating costs may prohibit you from warming your greenhouse above this temperature. Some species will succumb and desiccate if temperatures reach 95°F/35°C, so your greenhouse must have a cooling system, which can include fans, mist systems, air vents and shade cloth. Make sure you install "minimum-maximum" thermometers to measure the temperature in various areas of the greenhouse.

☐ **Light.** Sunlight can damage the leaves of your palms. Too much sun will bleach and burn their fronds; too little sun will produce weak palms with elongated leaves. Shade fabrics that block 50% will filter out unnecessary sun and create an optimal environment. The best systems for large greenhouses include solar-activated shades that open and close depending on the light intensity.

☐ **Humidity.** Palms prefer humidity levels at about 50-60%. Palms will show signs of leaf desiccation and insect infestation when humidity levels are too low. High humidity levels will produce algae, mold and fungal growth. Mist systems are the best way to improve the humidity in your greenhouse, but small greenhouse owners can simply spray the floor of the greenhouse with water to improve humidity. A humidistat will help monitor the humidity in your greenhouse.

☐ **Ventilation.** Good air circulation is crucial. Air should be constantly moving in the greenhouse to regulate the temperature and discourage the growth of fungus. Your greenhouse should include ventilators and fans placed throughout the structure. Smaller greenhouse owners can simply open the doors to circulate the air. Beware: open, unscreened doors and windows will allow unwanted insects and pests into your greenhouse.

☐ **Flooring.** The flooring in your greenhouse must be able to withstand dampness and water. Concrete, brick and gravel make good flooring materials. Choose a surface material that will not allow weeds to grow. Plastic weed fabric can also be used under benches to reduce the need for herbicides or hand weeding.

☐ **Benches.** Benches will improve your growing conditions by moderating the temperature and allowing air to circulate around your plants, inhibiting fungal growth. Small plants are easier to maintain when located at eye level. Plants on the ground may pick up soil pathogens and insects, and many local ordinances require nurseries to grow palms on benches for export.

Typical benches are made out of concrete blocks with wood planking. Metal and plastic benches typically are more expensive but provide better air circulation and sanitation.

☐ **Water.** Manually check the moisture content in your palms using a moisture probe or by feeling the soil below the surface. Unlike many other greenhouse plants, palms do not benefit from overhead mist systems, which can lead to bud rot. Manually apply water to the soil surface or use a drip system to inject water directly into the soil. Occasional drenches and sprays with water will leach soils of excess salts and prevent insect problems.

☐ **Fertilizer systems.** If you are using a drip system to water your palms, a fertilizer injector system is the ideal way to provide nutrients to your palms. Otherwise, apply soluble fertilizers or granular chemical fertilizers directly to the soil surface. Water your palms before you apply dry fertilizers to the soil.

☐ **Pest control.** Palms are typically pest-free, but greenhouse-grown palms are more susceptible to insects because of the lack of predators. Spider mites, aphids, scale and mealy bug are the most common culprits and can be controlled with common plant insecticides. To avoid most pest problems, follow these simple rules:

- Quarantine any new plants. Inspect any new introductions before they enter the greenhouse.
- Remove dead plant material from your palms. Insects hide in dead fronds.
- Keep your greenhouse weed-free.
- Screen windows, doors and vents if possible.

Some palms are easier to identify than others, including this bottle palm (*Hyophorbe lagenicaulis*) with its unusual swollen trunk.

Landscaping guides

Characteristics of 279 palm species

	Height			Spread			Light Needs			Water Needs			Minimum Temperature			Trunk Type		Leaf Type		Growth Rate			Salinity Tolerance			Key Traits	
	Under 25 ft./7 m.	25-50 ft./7-14 m.	Over 50 ft./14 m.	Under 10 ft./3 m.	10-20 ft./3-6 m.	Over 20 ft./6 m.	Shade	Partial Shade	Full Sun	Low	Medium	High	Below 20°F/-7°C	20-32°F/-7-0°C	Above 32°F/0°C	Single	Multiple	Palmate	Pinnate	Slow	Moderate	Fast	Low	Medium	High	Spines	Indoor
Acoelorrhaphe wrightii		●			●				●		●			●			●	●		●					●	●	
Acrocomia aculeata		●			●				●			●		●		●			●		●		●			●	
Adonidia merrillii		●			●			●			●			●		●			●		●				●		●
Aiphanes aculeata	●			●				●				●		●		●			●		●		●			●	●
Aiphanes minima	●			●				●				●		●		●			●		●		●			●	●
Allagoptera arenaria	●			●					●		●			●		●			●	●					●		
Allagoptera campestris	●			●					●		●			●		●			●	●					●		
Archontophoenix alexandrae		●			●				●			●		●		●			●			●	●				●
Archontophoenix cunninghamiana			●		●				●			●		●		●			●			●	●				●
A. cunninghamiana 'Illawarra'			●		●				●		●			●		●			●			●	●				●
Archontophoenix maxima		●			●			●				●		●		●			●			●	●				●
Archontophoenix myolensis		●			●			●				●		●		●			●			●	●				●
Archontophoenix purpurea		●			●			●			●			●		●			●			●	●				●
Archontophoenix tuckeri		●			●				●			●		●		●			●			●	●				●
Areca catechu		●			●				●			●			●	●			●			●		●			●
Areca concinna	●				●			●				●			●		●		●		●		●				●
Areca ipot	●			●				●				●		●		●			●			●	●				●
Areca triandra	●				●			●				●		●			●		●		●		●				●
Areca vestiaria	●				●			●				●			●		●		●		●		●				
Arenga engleri	●				●			●			●			●			●		●		●		●				●
Astrocaryum mexicanum	●			●				●			●			●		●			●	●			●			●	●
Attalea cohune		●				●			●			●		●		●			●	●			●				
Bactris gasipaes		●			●			●				●		●			●		●			●	●			●	
Beccariophoenix madagascariensis		●			●				●			●		●		●			●	●			●				
Bismarckia nobilis		●			●				●			●		●		●		●			●		●				
Borassus aethiopum			●			●			●			●		●		●		●		●			●				
Borassus flabellifer			●		●				●			●		●		●		●		●			●			●	
Brahea aculeata	●			●					●		●			●		●		●			●		●			●	
Brahea armata		●		●					●	●			●			●		●		●				●		●	
Brahea brandegeei		●			●				●	●			●			●		●		●			●			●	
Brahea decumbens	●				●				●	●				●			●	●		●			●			●	
Brahea dulcis	●			●					●		●			●		●		●		●			●			●	
Brahea edulis		●			●				●		●			●		●		●		●				●		●	
Brahea moorei	●			●				●	●		●		●			●		●		●			●			●	
Brahea pimo	●			●					●		●			●		●		●		●			●			●	
Burretiokentia hapala		●		●				●		●				●		●			●			●	●				
Burretiokentia viellardii		●		●				●		●				●		●			●			●	●				
Butia archeri	●			●					●	●				●		●			●	●			●				
Butia capitata	●			●					●		●		●			●			●	●				●			
Butia eriospatha	●				●				●	●			●			●			●	●			●				
Butia microspadix	●			●					●	●				●		●			●	●			●				

Characteristics of 279 palm species

	Height			Spread			Light Needs			Water Needs			Minimum Temperature			Trunk Type		Leaf Type		Growth Rate			Salinity Tolerance			Key Traits	
	Under 25 ft./7 m.	25-50 ft./7 -14 m.	Over 50 ft./14 m.	Under 10 ft./3 m.	10-20 ft./3-6 m.	Over 20 ft./6 m.	Shade	Partial Shade	Full Sun	Low	Medium	High	Below 20°F/-7°C	20-32°F/-7-0°C	Above 32°F/0°C	Single	Multiple	Palmate	Pinnate	Slow	Moderate	Fast	Low	Medium	High	Spines	Indoor
Butia paraguayensis	●				●				●	●				●		●			●	●			●				
Butia yatay		●			●				●		●		●			●			●	●			●				
Calamus australis	●			●			●					●		●			●		●	●			●			●	●
Calamus caryotoides	●			●			●					●	●				●		●	●			●			●	●
Calamus ciliaris	●			●			●					●		●		●			●	●			●			●	●
Calamus moti			●				●					●		●		●			●			●	●			●	●
Calamus muelleri		●		●			●					●			●		●		●			●	●			●	●
Calamus ornatus			●		●		●					●		●		●			●			●	●			●	●
Calamus radicalis		●		●			●					●		●		●			●			●	●			●	●
Calamus reyesianus		●		●			●					●		●		●			●			●	●			●	●
Calamus viminalis		●		●			●					●		●		●			●			●	●			●	●
Calyptrocalyx hollrungii	●			●				●				●		●			●		●				●				●
Carpentaria acuminata		●			●			●				●		●		●			●		●		●				●
Carpoxylon macrospermum			●		●			●				●		●		●			●	●			●				
Caryota maxima		●		●				●				●		●		●			●			●	●				●
Caryota mitis	●				●			●				●		●			●		●			●	●				●
Caryota no			●		●			●				●		●		●			●			●	●				●
Caryota obtusa			●		●			●				●		●		●			●			●	●				●
Caryota ochlandra		●			●			●				●		●		●			●			●	●				●
Caryota rumphiana		●			●			●				●		●		●			●			●	●				●
Caryota urens		●			●				●			●		●		●			●			●	●				●
Ceroxylon alpinum		●				●		●				●		●		●			●	●			●				
Ceroxylon quindiuense			●			●		●				●		●		●			●	●			●				
Ceroxylon ventricosum			●		●			●				●		●		●			●	●			●				
Chamaedorea cataractarum	●			●			●					●		●			●		●	●			●				●
Chamaedorea costaricana	●			●				●				●		●			●		●			●	●				●
Chamaedorea elegans	●			●				●				●				●			●	●			●				●
Chamaedorea ernestii-augusti	●			●			●					●		●		●			●	●			●				●
Chamaedorea fragrans	●			●			●					●		●			●		●	●			●				●
Chamaedorea glaucifolia	●			●			●					●		●		●			●	●			●				●
Chamaedorea hooperiana	●			●			●					●		●			●		●	●			●				●
Chamaedorea klotzschiana	●			●			●					●		●		●			●	●			●				●
Chamaedorea metallica	●			●			●					●		●		●			●	●			●				●
Chamaedorea microspadix	●			●				●				●		●			●		●		●		●				●
Chamaedorea plumosa	●			●					●			●		●		●			●			●	●				●
Chamaedorea pochutlensis	●			●			●					●		●			●		●			●	●				●
Chamaedorea radicalis	●			●				●				●		●		●			●	●			●				●
Chamaedorea seifrizii	●			●				●				●		●			●	●		●			●				●
Chamaedorea stolonifera	●			●				●				●		●			●		●	●			●				●
Chamaedorea woodsoniana	●			●				●				●		●		●			●	●			●				●
Chamaerops humilis	●				●				●		●		●				●	●		●				●		●	●

Characteristics of 279 palm species

	Height			Spread			Light Needs			Water Needs			Minimum Temperature			Trunk Type		Leaf Type		Growth Rate			Salinity Tolerance			Key Traits	
	Under 25 ft./7 m.	25-50 ft./7 -14 m.	Over 50 ft./14 m.	Under 10 ft./3 m.	10-20 ft./3-6 m.	Over 20 ft./6 m.	Shade	Partial Shade	Full Sun	Low	Medium	High	Below 20°F/-7°C	20-32°F/-7-0°C	Above 32°F/0°C	Single	Multiple	Palmate	Pinnate	Slow	Moderate	Fast	Low	Medium	High	Spines	Indoor
Chambeyronia lepidota	●			●				●				●		●		●				●			●				
Chambeyronia macrocarpa	●				●			●				●		●		●			●			●	●				●
Clinostigma exorrhizum		●			●			●				●			●	●			●		●		●				
Clinostigma gronophyllum	●				●			●				●			●	●			●		●		●				
Clinostigma harlandii	●				●			●				●			●	●			●	●			●				
Clinostigma ponapense			●		●			●				●			●	●			●		●		●				
Clinostigma samoense		●			●			●				●			●	●			●		●		●				
Clinostigma savoryanum	●				●			●				●			●	●			●	●			●				
Coccothrinax argentata	●			●					●		●			●		●		●		●					●		
Coccothrinax argentea	●			●					●		●			●		●		●		●					●		
Coccothrinax barbadensis	●			●					●		●			●		●		●		●					●		
Coccothrinax borhidiana		●			●				●		●			●		●		●		●					●		
Coccothrinax crinita	●				●				●		●			●		●		●		●					●		
Coccothrinax cupularis	●			●					●		●			●		●		●		●					●		
Coccothrinax ekmanii	●			●					●		●			●		●		●		●					●		
Coccothrinax fragrans		●		●					●		●			●		●		●		●					●		
Coccothrinax hioramii		●			●				●		●			●		●		●		●					●		
Coccothrinax inaguensis	●			●							●					●									●		
Coccothrinax jamaicensis		●		●					●		●			●		●		●		●					●		
Coccothrinax miraguama	●				●				●		●			●		●		●		●					●		
Coccothrinax munizii		●		●					●		●			●		●		●		●					●		
Coccothrinax proctorii		●		●					●		●			●		●		●		●					●		
Coccothrinax spissa		●			●				●		●			●		●		●		●					●		
Cocos nucifera			●		●				●			●			●	●			●			●			●		
Copernicia alba		●		●					●		●			●		●		●			●				●		
Copernicia baileyana		●			●				●		●			●		●		●		●					●	●	
Copernicia glabrescens		●		●					●		●		●			●		●		●				●		●	
Copernicia hospita	●			●					●		●		●			●		●		●					●	●	
Copernicia macroglossa	●				●				●		●			●		●		●		●					●	●	
Copernicia prunifera		●			●				●		●			●		●		●		●					●	●	
Corypha umbraculifera			●		●				●			●		●		●		●			●		●			●	
Corypha utan			●		●				●			●		●		●		●			●		●			●	
Cryosophila guagara		●		●			●					●				●		●		●			●			●	
Cryosophila stauracantha		●		●				●				●				●		●		●			●			●	
Cryosophila warscewiczii	●			●				●				●		●		●		●		●			●			●	
Cryosophila williamsii	●			●				●				●		●		●		●		●			●			●	
Cyrtostachys renda		●			●				●			●				●	●		●	●			●				
Dictyosperma album		●			●			●				●		●		●			●		●			●			●
Dypsis ambositrae	●				●			●			●			●		●			●	●			●				
Dypsis baronii	●			●				●			●			●			●		●	●			●				
Dypsis cabadae	●				●			●			●			●			●		●		●		●				●

Characteristics of 279 palm species

	Height			Spread			Light Needs			Water Needs			Minimum Temperature			Trunk Type		Leaf Type		Growth Rate			Salinity Tolerance			Key Traits	
	Under 25 ft./7 m.	25-50 ft./7 -14 m.	Over 50 ft./14 m.	Under 10 ft./3 m.	10-20 ft./3-6 m.	Over 20 ft./6 m.	Shade	Partial Shade	Full Sun	Low	Medium	High	Below 20°F/-7°C	20-32°F/-7-0°C	Above 32°F/0°C	Single	Multiple	Palmate	Pinnate	Slow	Moderate	Fast	Low	Medium	High	Spines	Indoor
Dypsis decaryi		●			●			●			●		●			●			●		●		●				
Dypsis decipiens		●			●				●		●			●		●	●		●		●		●				
Dypsis lastelliana			●		●			●			●			●		●			●	●			●				
Dypsis leptocheilos			●		●			●			●			●		●			●	●			●				
Dypsis lutescens	●				●			●			●			●			●		●		●		●				●
Dypsis madagascariensis		●			●			●			●			●		●			●		●		●				
Elaeis guineensis		●			●				●		●			●		●			●		●			●		●	
Elaeis oleifera	●				●				●		●			●		●			●		●		●				
Euterpe edulis	●				●			●				●		●		●			●		●		●				●
Euterpe oleracea		●			●			●				●		●			●	●			●		●				●
Gastrococos crispa			●		●				●			●		●		●			●		●		●			●	
Gaussia attenuata	●				●			●			●			●		●			●	●			●				
Gaussia maya	●				●			●			●			●		●			●	●			●				
Gaussia princeps	●				●			●			●			●		●			●	●			●				
Guihaia argyrata	●			●				●			●		●				●	●		●			●				●
Hedyscepe canterburyana		●			●			●			●			●		●			●	●				●			●
Heterospathe elata		●			●			●				●		●		●			●		●		●				●
Heterospathe negrosensis	●			●				●				●		●		●			●	●			●				●
Howea belmoreana		●			●			●			●			●		●			●	●				●			●
Howea forsteriana		●			●			●			●			●		●			●	●				●			●
Hydriastele microspadix		●			●			●				●		●			●		●	●			●				
Hyophorbe indica	●				●			●				●		●		●			●	●				●			●
Hyophorbe lagenicaulis	●			●					●			●			●	●			●	●				●			●
Hyophorbe verschaffeltii	●			●					●			●		●		●			●	●				●			●
Hyphaene coriacea	●				●				●	●				●		●		●		●				●			
Hyphaene dichotoma			●			●			●	●				●		●		●		●				●			
Hyphaene petersiana		●				●			●	●				●		●		●		●				●			
Hyphaene thebaica		●			●				●	●				●		●		●		●				●			
Johannesteijsmannia altifrons	●				●		●					●		●		●		●		●			●				
Jubaea chilensis			●			●			●		●		●			●			●	●			●				
Jubaeopsis caffra	●					●			●			●		●			●		●	●			●				
Kentiopsis oliviformis			●		●			●			●			●		●			●	●			●				
Laccospadix australasicus	●				●			●				●		●			●		●		●		●				●
Latania loddigesii		●			●				●		●			●		●		●		●				●			
Latania lontaroides		●			●				●		●			●		●		●		●			●				
Latania verschaffeltii		●			●				●		●			●		●		●		●			●				
Lepidorrhachis mooreana	●			●				●				●		●		●			●	●				●			
Licuala cordata	●			●			●					●		●		●		●		●			●			●	
Licuala grandis	●			●			●					●		●		●		●		●				●		●	●
Licuala orbicularis	●			●			●					●		●		●		●		●			●			●	
Licuala peltata	●			●			●					●		●		●		●			●			●		●	

Characteristics of 279 palm species

	Height			Spread			Light Needs			Water Needs			Minimum Temperature			Trunk Type		Leaf Type		Growth Rate			Salinity Tolerance			Key Traits	
	Under 25 ft./7 m.	25-50 ft./7 -14 m.	Over 50 ft./14 m.	Under 10 ft./3 m.	10-20 ft./3-6 m.	Over 20 ft./6 m.	Shade	Partial Shade	Full Sun	Low	Medium	High	Below 20°F/-7°C	20-32°F/-7-0°C	Above 32°F/0°C	Single	Multiple	Palmate	Pinnate	Slow	Moderate	Fast	Low	Medium	High	Spines	Indoor
Licuala ramsayi		●		●			●					●		●		●		●		●				●		●	
Licuala spinosa	●			●			●					●		●			●	●		●				●		●	
Linospadix monostachya	●			●				●				●		●		●			●	●			●				●
Livistona australis		●		●					●		●		●			●		●		●				●		●	
Livistona chinensis		●			●				●		●		●			●		●			●			●		●	
Livistona decora		●		●					●		●			●		●		●			●			●		●	
Livistona drudei		●		●					●		●			●		●		●			●			●		●	
Livistona inermis	●			●					●		●			●		●		●		●				●		●	
Livistona jenkinsiana		●			●				●		●			●		●		●		●				●		●	
Livistona mariae		●			●				●		●			●		●		●			●			●		●	
Livistona muelleri	●			●					●		●			●		●		●		●				●		●	●
Livistona saribus		●			●				●		●	●		●		●		●			●			●		●	●
Lodoicea maldivica			●			●			●			●			●	●		●		●			●				
Lytocaryum hoehnei	●			●				●				●		●		●			●	●			●				●
Lytocaryum weddellianum	●			●				●				●		●		●			●	●			●				●
Medemia argun		●				●			●	●			●			●		●		●					●		
Metroxylon amicarum			●			●		●				●			●	●			●			●	●			●	
Metroxylon sagu			●			●		●				●			●		●		●			●	●			●	
Metroxylon vitiense		●				●		●				●			●	●			●			●	●			●	
Metroxylon warburgii	●					●		●				●			●	●			●			●	●			●	
Nannorrhops ritchiana	●				●				●		●		●				●	●		●				●			
Neoveitchia storckii		●			●				●			●		●		●			●			●	●				
Normanbya normanbyi	●				●			●				●		●		●			●	●			●				●
Nypa fruticans	●				●				●			●		●			●		●		●			●			
Oncosperma tigillarium			●		●			●				●		●			●		●		●				●	●	
Parajubaea cocoides			●		●				●		●			●		●			●		●		●				
Parajubaea torallyi		●			●				●		●			●		●			●	●			●				
Phoenix canariensis			●			●			●	●				●		●			●			●		●		●	
Phoenix dactylifera			●		●				●	●			●			●			●		●				●	●	
Phoenix loureiroi	●				●				●		●			●		●	●		●	●				●		●	
Phoenix pusilla	●				●				●		●			●		●			●	●			●			●	
Phoenix reclinata		●				●			●		●			●			●		●		●		●			●	
Phoenix roebelenii	●			●				●			●			●		●	●		●	●			●			●	●
Phoenix rupicola		●			●				●		●			●		●			●		●			●		●	●
Phoenix sylvestris		●			●				●		●			●		●			●		●			●		●	
Phoenix theophrasti		●			●				●		●		●				●		●		●				●	●	
Pinanga caesia	●				●			●				●		●		●			●	●			●				
Pinanga coronata	●			●				●				●		●			●		●	●			●				●
Pinanga javana	●			●				●				●		●		●			●	●			●				●
Pinanga philippinensis	●			●				●				●		●			●		●	●			●				●
Pritchardia affinis		●			●			●			●			●		●		●				●	●				

Characteristics of 279 palm species

	Height			Spread			Light Needs			Water Needs			Minimum Temperature			Trunk Type		Leaf Type		Growth Rate			Salinity Tolerance			Key Traits	
	Under 25 ft./7 m.	25-50 ft./7 -14 m.	Over 50 ft./14 m.	Under 10 ft./3 m.	10-20 ft./3-6 m.	Over 20 ft./6 m.	Shade	Partial Shade	Full Sun	Low	Medium	High	Below 20°F/-7°C	20-32°F/-7-0°C	Above 32°F/0°C	Single	Multiple	Palmate	Pinnate	Slow	Moderate	Fast	Low	Medium	High	Spines	Indoor
Pritchardia arecina		●			●				●		●			●		●		●			●		●				
Pritchardia beccariana		●			●			●			●			●		●		●		●			●				
Pritchardia hillebrandii		●			●				●		●			●		●		●		●			●				
Pritchardia kaalae		●			●			●			●			●		●		●		●			●				
Pritchardia lanaiensis		●			●				●		●			●		●		●			●		●				
Pritchardia lowreyana	●			●					●		●			●		●		●		●			●				
Pritchardia martii	●			●				●			●			●		●		●		●			●				
Pritchardia pacifica		●			●				●		●				●	●		●			●			●			
Pritchardia remota	●			●				●			●			●		●		●			●		●				
Pritchardia thurstonii		●			●				●		●				●	●		●			●		●				
Pseudophoenix sargentii	●				●				●	●				●		●			●	●					●		
Pseudophoenix vinifera	●				●				●	●				●		●			●	●					●		
Ptychosperma elegans			●		●			●				●		●		●			●	●			●				●
Ptychosperma macarthurii		●		●				●				●		●			●		●	●			●				●
Raphia farinifera		●				●			●			●		●			●		●		●		●				
Raphia hookeri		●				●			●			●		●			●		●		●		●				
Ravenea glauca	●				●			●				●		●		●			●	●				●			
Ravenea hildebrandtii	●			●				●				●		●		●			●	●			●				
Ravenea rivularis		●			●				●			●		●		●			●			●		●			
Rhapidophyllum hystrix	●				●			●	●		●		●				●	●		●				●		●	
Rhapis excelsa	●			●				●			●			●			●	●			●			●			●
Rhapis humilis	●			●				●			●			●			●	●		●				●			●
Rhapis subtilis	●			●				●			●			●			●	●		●			●				●
Rhopaloblaste elegans		●			●			●				●		●		●		●			●		●				
Rhopalostylis baueri		●			●			●				●		●		●			●	●			●				●
Rhopalostylis cheesemanii		●			●			●				●		●		●			●	●			●				●
Rhopalostylis sapida		●			●			●				●		●		●			●	●			●				●
Roystonea altissima		●			●				●			●		●		●			●			●		●			
Roystonea borinquena		●			●				●			●		●		●			●		●			●			
Roystonea oleracea			●			●			●			●		●		●			●		●			●			
Roystonea princeps		●			●				●			●		●		●			●		●			●			
Roystonea regia			●		●				●			●		●		●			●		●			●			
Sabal bermudana	●				●				●		●		●			●		●		●				●			
Sabal causiarum		●			●				●		●		●			●		●		●				●			
Sabal domingensis		●			●				●		●			●		●		●		●				●			
Sabal etonia	●			●					●		●			●		●		●		●				●			
Sabal mexicana			●		●				●		●				●	●		●		●				●			
Sabal minor	●			●					●		●		●			●		●		●				●			
Sabal palmetto		●			●				●		●		●			●		●		●				●			
Sabal 'Riverside'	●				●				●		●		●			●		●		●				●			
Sabal rosei			●		●				●		●		●			●		●		●				●			

Characteristics of 279 palm species

	Height			Spread			Light Needs			Water Needs			Minimum Temperature			Trunk Type		Leaf Type		Growth Rate			Salinity Tolerance			Key Traits	
	Under 25 ft./7 m.	25-50 ft./7 -14 m.	Over 50 ft./14 m.	Under 10 ft./3 m.	10-20 ft./3-6 m.	Over 20 ft./6 m.	Shade	Partial Shade	Full Sun	Low	Medium	High	Below 20°F/-7°C	20-32°F/-7-0°C	Above 32°F/0°C	Single	Multiple	Palmate	Pinnate	Slow	Moderate	Fast	Low	Medium	High	Spines	Indoor
Sabal uresana		●			●				●		●			●		●		●		●				●			
Satakentia liukiuensis		●			●			●				●		●		●			●		●			●			
Schippia concolor	●			●				●			●			●		●		●		●				●			
Serenoa repens	●				●				●		●		●				●	●		●					●	●	
Syagrus amara		●			●				●		●			●		●			●		●			●			
Syagrus cearensis		●				●			●		●			●			●	●			●			●			
Syagrus comosa	●				●				●		●			●		●	●		●	●				●			
Syagrus coronata	●				●				●		●			●		●			●	●				●			
Syagrus pseudococos		●			●				●		●			●		●			●	●				●			
Syagrus romanzoffiana			●		●				●		●			●		●			●			●		●			
Syagrus sancona			●		●				●		●			●		●			●		●			●			
Syagrus schizophylla	●				●				●		●			●		●			●	●				●			
Thrinax excelsa	●			●					●		●			●		●		●		●				●			
Thrinax morrisii		●		●				●			●			●		●		●		●					●		●
Thrinax parviflora	●			●				●			●			●		●		●		●				●			
Thrinax radiata	●			●				●			●			●		●		●		●				●			
Trachycarpus fortunei		●		●					●		●		●			●		●			●			●			
Trachycarpus latisectus		●		●					●		●		●			●		●			●		●				
Trachycarpus martianus		●		●					●		●		●			●		●		●			●				
Trachycarpus nanus	●			●					●		●		●			●		●		●			●				
Trachycarpus takil		●		●					●		●		●			●		●			●		●				
Trithrinax brasiliensis		●			●				●		●		●			●	●	●		●			●			●	
Trithrinax campestris	●				●				●		●		●				●	●		●			●			●	
Veitchia arecina		●			●			●				●		●		●			●		●			●			●
Veitchia filifera		●			●			●				●			●	●			●			●		●			●
Veitchia joannis			●		●			●				●		●		●			●			●		●			●
Veitchia winin		●			●			●				●			●	●			●			●		●			
Verschaffeltia splendida		●			●				●			●			●	●			●		●		●			●	
Washingtonia filifera			●		●				●		●		●			●		●				●		●		●	
Washingtonia robusta			●		●				●		●			●		●		●				●		●		●	
Welfia regia			●		●				●			●		●		●			●	●			●				
Wodyetia bifurcata	●			●					●		●			●		●			●	●			●				
Zombia antillarum	●			●				●			●			●			●	●		●				●		●	

Sabal palmetto

PALMS FOR
Coastal areas

Palms planted in coastal areas must weather harsh winds and continual sprays of salt-laden air. Many palms can adapt to these conditions, and some will thrive here. Choose palms that are native to coastal areas around the world. Alternatively, many desert palms can also do well in coastal areas, able to withstand poor saline soils. Some palms, including the Canary Island date palm (*Phoenix canariensis)*, can tolerate sea spray, but cannot live in brackish water. On a regular basis, provide your palms with deep soaks of fresh water. Hose down foliage to remove salt buildup on the fronds and trunks.

Botanical Name	Common Name
Acoelorrhaphe wrightii	paurotis palm, silver saw palm, Everglades palm
Adonidia merrillii	Manila palm, Christmas palm
Allagoptera arenaria	sand palm
Allagoptera campestris	sand palm
Areca catechu	betel nut palm
Brahea armata	Mexican blue fan palm, blue hesper palm
Brahea edulis	Guadalupe palm
Butia capitata	pindo palm, jelly palm
Chamaerops humilis	Mediterranean fan palm
Coccothrinax species	
Cocos nucifera	coconut palm
Copernicia alba	caranday palm
Copernicia baileyana	Bailey fan palm
Copernicia glabrescens	guano palm
Copernicia species	
Dictyosperma album	hurricane palm, princess palm
Elaeis guineensis	African oil palm
Hedyscepe canterburyana	umbrella palm, big mountain palm
Howea species	

Botanical Name	Common Name
Hyophorbe species	
Hyphaene species	
Latania loddigesii	blue Latan palm
Lepidorrhachis mooreana	little mountain palm
Licuala species	
Livistona species	
Medemia argun	
Nannorrhops ritchiana	Mazari palm
Nypa fruticans	mangrove palm
Oncosperma tigillarium	nibung palm
Phoenix canariensis	Canary Island date palm
Phoenix dactylifera	edible date palm
Phoenix loureiroi	voiavoia palm
Phoenix rupicola	cliff date palm
Phoenix sylvestris	silver date palm, toddy palm, wild date palm
Phoenix theophrasti	Cretan date palm
Pritchardia pacifica	Fiji fan palm
Pseudophoenix sargentii	cherry palm
Pseudophoenix vinifera	buccaneer palm, wine palm
Ravenea glauca	
Ravenea rivularis	majesty palm

Botanical Name	Common Name
Rhapidophyllum hystrix	needle palm
Rhapis excelsa	lady palm
Rhapis humilis	slender lady palm
Roystonea altissima	Jamaican royal palm, mountain cabbage royal
Roystonea borinquena	Puerto Rican royal palm
Roystonea oleracea	South American, Venezuelan royal palm
Roystonea princeps	Jamaican royal palm
Roystonea regia	Cuban royal palm
Sabal species	
Satakentia liukiuensis	satake palm
Schippia concolor	silver pimento palm, mountain pimento palm
Serenoa repens	saw palmetto
Syagrus species	
Thrinax species	
Trachycarpus fortunei	Chinese windmill palm, chusan palm
Trithrinax brasiliensis	spiny fiber palm
Veitchia species	
Washingtonia species	
Zombia antillarum	zombie palm

Chamaedorea elegans

PALMS FOR
Indoors

There is no such thing as an indoor palm, but if you can emulate a palm's native environment, you can grow palms indoors with confidence. Temperature, light, moisture and humidity are the key factors you must control to successfully recreate the palm's natural environment. The following palms are more tolerant than others of indoor conditions. Some, like the *Archontophoenix* genus, will outgrow most indoor areas but make good houseplants as juvenile plants.

Botanical Name	Common Name
Adonidia merrillii	Manila palm, Christmas palm
Aiphanes aculeata	gru gru palm, ruffle palm, coyure palm
Aiphanes minima	gru gru palm, ruffle palm, coyure palm
Archontophoenix species	
Areca catechu	betel nut palm
Areca concinna	
Areca ipot	bunga ipot, highland betel nut palm
Areca triandra	
Arenga engleri	dwarf sugar palm, Formosa palm
Astrocaryum mexicanum	Chocho, Lancetilla, Chichon
Calamus species	
Calyptrocalyx hollrungii	
Carpentaria acuminata	carpentaria palm
Caryota species	

Botanical Name	Common Name
Chamaedorea species	
Chamaerops humilis	Mediterranean fan palm
Chambeyronia macrocarpa	flamethrower
Dictyosperma album	hurricane palm, princess palm
Dypsis cabadae	cabada palm
Dypsis lutescens	areca palm
Euterpe species	
Guihaia argyrata	Guangxi palm
Hedyscepe canterburyana	umbrella palm, big mountain palm
Heterospathe species	
Howea species	
Hyophorbe species	
Laccospadix australasicus	Atherton palm
Licuala grandis	licuala palm

Botanical Name	Common Name
Linospadix monostachya	walking stick palm
Livistona muelleri	dwarf fan palm
Livistona saribus	taraw palm
Lytocaryum hoehnei	
Lytocaryum weddellianum	Victorian parlor palm
Normanbya normanbyi	black palm
Phoenix roebelenii	pygmy date palm
Phoenix rupicola	cliff date palm
Pinanga coronata	Coronata palm, ivory cane palm
Pinanga javana	
Pinanga philippinensis	
Ptychosperma species	
Rhapis species	
Rhopalostylis species	
Thrinax morrissii	brittle thatch palm
Veitchia species	

Phoenix canariensis

PALMS FOR
Arid areas

The following list of palms includes species that will grow in semi-arid and desert conditions. These palms are wind and drought tolerant, able to survive periods with no irrigation or rainfall.

Botanical Name	Common Name
Brahea species	
Butia species	
Chamaerops humilis	Mediterranean fan palm
Coccothrinax species	
Copernicia macroglossa	Cuban petticoat palm

Botanical Name	Common Name
Corypha utan	talipot palm, gebang palm, buri palm
Dypsis ambrositrae	
Dypsis decaryi	triangle palm
Hyphaene species	
Medemia argun	

Botanical Name	Common Name
Phoenix canariensis	Canary Island date palm
Phoenix dactylifera	edible date palm
Pseudophoenix species	
Serenoa repens	saw palmetto
Thrinax morrissii	brittle thatch palm

Phoenix roebelenii

PALMS FOR

Small gardens

Hundreds of palms make excellent landscape specimens for small gardens or areas with limited space. Below are several species that typically mature to a height of less than 25 ft./7 m. tall. The ultimate height of these palms will vary based on your climate and growing conditions. Palms in their native habitat and areas that emulate their natural growing conditions will typically grow taller than palms grown in sub-optimal conditions. For example, tropical palms will typically grow taller in Hawaii than in the subtropical climate of California.

Full Sun

Botanical Name	Common Name
Allagoptera arenaria	sand palm
Allagoptera campestris	sand palm
Brahea aculeata	Sinaloan hesper palm
Brahea dulcis	
Brahea decumbens	Sierra Madre palm, Mexican dwarf blue palm
Brahea pimo	
Butia species	
Chamaerops humilis	Mediterranean fan palm
Coccothrinax species	
Copernicia hospita	hospita palm
Copernicia macroglossa	Cuban petticoat palm
Elaeis oleifera	American oil palm
Hyophorbe species	
Hyphaene coriacea	ilala palm
Jubaeopsis caffra	pondoland palm
Nannorrhops ritchiana	Mazari palm
Nypa fruticans	mangrove palm
Phoenix loureiroi	voiavoia palm
Phoenix pusilla	Ceylon date palm
Pseudophoenix species	
Sabal bermudana	Bermuda palm
Sabal etonia	scrub palmetto
Serenoa repens	saw palmetto
Syagrus coronata	licury palm
Syagrus comosa	babao
Syagrus schizophylla	aricury palm
Thrinax excelsa	broad thatch palm
Trachycarpus nanus	Yunnan dwarf palm
Trithrinax campestris	Caranday palm or Coronday palm
Wodyetia bifurcata	foxtail palm

Partial Shade

Botanical Name	Common Name
Aiphanes species	
Areca concinna	
Areca ipot	bunga ipot, highland betel nut palm
Areca vestiaria	orange collar palm
Arenga engleri	dwarf sugar palm, Formosa palm
Astrocaryum mexicanum	Chocho, Lancetilla, Chichon
Calamus caryotoides	fishtail lawyer's cane
Caryota mitis	clustering fishtail palm
Chamaedorea species	
Chambeyronia species	
Clinostigma gronophyllum	
Clinostigma harlandii	
Clinostigma savoryanum	
Dypsis baronii	baron palm
Dypsis cabadae	cabada palm
Dypsis lastelliana	redneck palm
Dypsis lutescens	areca palm, butterfly palm
Euterpe species	
Gaussia species	
Guihaia argyrata	Guangxi palm
Heterospathe negrosensis	
Hyophorbe indica	
Laccospadix australasicus	Atherton palm
Lepidorrhachis mooreana	little mountain palm
Licuala peltata	elegant palm
Livistona inermis	wispy fan palm
Livistona muelleri	dwarf fan palm
Normanbya normanbyi	black palm
Phoenix roebelenii	pygmy date palm
Pinanga javana	
Pritchardia martii	dwarf Hawaiian fan palm
Pritchardia minor	
Pritchardia remota	Nihoa palm
Ptychosperma macarthurii	Macarthur palm
Ravenea glauca	
Ravenea hildebrandtii	
Rhapidophyllum hystrix	needle palm
Rhapis species	
Sabal minor	dwarf blue palmetto
Schippia concolor	silver pimento palm, mountain pimento palm
Thrinax parviflora	broom palm
Zombia antillarum	zombie palm

Full Shade

Botanical Name	Common Name
Areca triandra	
Calamus australis	lawyer cane, hairy Mary
Chamaedorea species	
Johannesteijsmannia altifrons	joey palm, diamond joey palm, umbrella leaf palm
Licuala species	
Linospadix monostachya	walking stick palm
Lytocaryum species	
Pinanga coronata	Coronata palm, ivory cane palm
Pinanga philippinensis	

Arenga engleri

PALMS FOR

Screens and hedges

Many palms make ideal screens and can be used to hedge a property or to screen out an unsightly area. Choose clumping palms that maintain their fronds from the ground up. Several palms, like *Phoenix theophrasti*, will make large-scale screens, provided you maintain the trunks and fronds. Others, like *Rhapis excelsa*, make ideal small-scale screens to hide utilities, air conditioners, trash facilities and more.

Botanical Name	Common Name
Acoelorrhaphe wrightii	paurotis palm, Everglades palm, silver saw palmetto
Areca concinna	
Areca triandra	
Areca vestiaria	orange collar palm
Arenga engleri	dwarf sugar palm, Formosa palm
Bactris gasipaes	peach palm, peijibaye
Calyptrocalyx hollrungii	
Caryota mitis	clustering fishtail palm
Chamaedorea cataractarum	cat palm, cascade palm
Chamaedorea costaricana	bamboo palm
Chamaedorea fragrans	sangapilla
Chamaedorea hooperiana	

Botanical Name	Common Name
Chamaedorea microspadix	hardy bamboo palm
Chamaedorea pochutlensis	cannellila
Chamaerops humilis	Mediterranean fan palm
Cyrtostachys renda	sealing wax palm, lipstick palm
Dypsis baronii	baron palm
Dypsis cabadae	cabada palm
Dypsis lutescens	areca palm, butterfly palm
Euterpe oleracea	Assai palm
Guihaia argyrata	Guangxi palm
Hydriastele microspadix	
Jubaeopsis caffra	Pondoland palm
Metroxylon sagu	sago palm
Nannorrhops ritchiana	Mazari palm

Botanical Name	Common Name
Oncosperma tigillarium	nibung palm
Phoenix reclinata	Senegal date palm
Phoenix theophrasti	Cretan date palm
Pinanga coronata	Coronata palm, ivory cane palm
Pinanga philippinensis	
Ptychosperma macarthurii	Macarthur palm
Raphia farinifera	raffia palm
Raphia hookerii	raffia palm
Rhapidophyllum hystrix	needle palm
Rhapis excelsa	lady palm
Rhapis humilis	slender lady palm
Rhapis subtilis	dwarf lady palm
Serenoa repens	saw palmetto
Trithrinax campestris	Caranday palm
Zombia antillarum	zombie palm

Livistona spines

PALMS WITH

Spines

Some of the most beautiful species of palms are armed with spines, either on their trunks or their petioles and leaves. Use caution when planting spiny palms near areas where people may be injured, such as walkways, patios and pools. In some cases, spines may disappear as the palm matures. Some palms, including the *Phoenix* genus, may grow tall enough that their spines are out of reach.

Botanical Name	Common Name
Acoelorrhaphe wrightii	paurotis palm, Everglades palm, silver saw palmetto
Acrocomia aculeata	gru gru palm
Aiphanes aculeata	gru gru palm, ruffle palm, coyure palm
Aiphanes minima	gru gru palm, ruffle palm, coyure palm
Astrocaryum mexicanum	Chocho, Lancetilla, Chichon
Bactris gasipaes	peach palm, peijibaye
Borassus flabellifer	Palmyra palm, toddy palm, Lontar palm

Botanical Name	Common Name
Brahea species	
Calamus species	
Chamaerops humilis	Mediterranean fan palm
Copernicia species	
Corypha species	
Cryosophila species	
Elaeis guineensis	African oil palm
Gastrococos crispa	Cuban belly palm
Licuala species	

Botanical Name	Common Name
Livistona species	
Metroxylon species	
Oncosperma tigillarium	nibung palm
Phoenix species	
Rhapidophyllum hystrix	needle palm
Serenoa repens	saw palmetto
Trithrinax species	
Verschaffeltia splendida	Seychelles stilt palm
Washingtonia species	
Zombia antillarum	zombie palm

Caryota urens

PALMS THAT

Grow fast

Fast is a relative word when describing the growth rate of palms. Depending on your environment, some palms will grow faster than others. For example, palms from dry climate areas will grow faster in California than tropical rainforest palms. The Canary Island date palm (*Phoenix canariensis*) will grow faster in California than it does in more humid Florida. Similarly, a Cuban royal palm will grow faster in Miami than it does in Los Angeles.

As a rule, most palms are relatively slow growing, but the following list includes palms that typically grow faster than most.

Botanical Name	Common Name
Archontophoenix species	Alexandra palm, king palm, picabeen
Areca catechu	betel nut palm
Areca ipot	bunga ipot, highland betel nut palm
Bactris gasipaes	peach palm, peijibaye
Burretiokentia species	
Calamus moti	yellow lawyer cane
Calamus species	
Caryota species	
Chamaedorea costaricana	bamboo palm
Chamaedorea plumosa	
Chamaedorea pochutlensis	cannellila
Chambeyronia macrocarpa	flamethrower
Cocos nucifera	coconut palm
Metroxylon species	
Neoveitchia storckii	unleito
Phoenix canariensis	Canary Island date palm
Pritchardia affinis	Kona fan palm
Ravenea rivularis	majesty palm
Roystonea altissima	Jamaican royal palm, mountain cabbage royal
Syagrus romanzoffiana	queen palm
Veitchia species	
Washingtonia species	

Trachycarpus fortunei

PALMS FOR

Cold climates

A surprising number of palms can sustain temperatures well below freezing. The following list includes palms that will survive temperatures below 20°F/ -7°C for short periods of time. This selection is based on the latest information available to palm growers. If you live in a cold-climate area, this list will provide a good starting point for your selection. Humidity, moisture, wind and other environmental factors affect the cold hardiness of palms. The age of the palm and the duration of the cold season will also affect your success. The seasonal checklist in this book contains strategies to improve the cold hardiness of your palms.

Botanical Name	Common Name
Brahea armata	Mexican blue fan palm, blue hesper palm
Brahea brandegeei	San Jose hesper palm, palma de tlaco, palma negra
Brahea moorei	dwarf rock palm, powder palm
Butia capitata	pindo palm, jelly palm
Butia eriospatha	wooly butia
Butia yatay	
Calamus caryotoides	fishtail lawyer's cane
Chamaerops humilis	Mediterranean fan palm
Copernicia glabrescens	guano palm
Copernicia hospita	hospita palm
Dypis decaryi	triangle palm
Guihaia argyrata	Guangzi palm, Chinese needle palm, dainty lady palm
Jubaea chilensis	Chilean wine palm, coquito palm, honey palm
Livistona australis	Australian fan palm
Livistona chinensis	Chinese fan palm, Chinese fountain palm
Medemia argun	
Nannorrhops ritchiana	Mazari palm
Phoenix dactylifera	edible date palm
Phoenix theophrasti	Cretan date palm
Rhapidophyllum hystrix	needle palm
Sabal bermudana	Bermuda palm
Sabal causiarum	Puerto Rican hat palm, Panama hat palm
Sabal palmetto	palmetto palm
Sabal 'Riverside'	
Sabal rosei	savannah palmetto
Serenoa repens	saw palmetto
Trachycarpus species	
Trithrinax species	
Washingtonia filifera	California fan palm

Chamaedorea radicalis

PALMS FOR

Shade

A wide variety of palms thrive in shade, particularly when they are young. Many palm species germinate in the shade of older trees, becoming more sun tolerant as they reach for the canopy. The following list will grow well in areas that are shaded but receive several hours of filtered sunlight. In foggy coastal areas, these palms can be planted in direct sunlight. The full shade list includes palms that can live indefinitely in low light.

Botanical Name	Common Name
Adonidia merrillii	Manila palm, Christmas palm
Aiphanes aculeata	gru gru palm, ruffle palm, coyure palm
Aiphanes minima	gru gru palm, ruffle palm, coyure palm
Archontophoenix maxima	Walsh River king palm
Archontophoenix myolensis	Myola palm
Archontophoenix purpurea	Mount Lewis king palm
Areca concinna	
Areca ipot	bunga ipot, highland betel nut palm
Areca vestiaria	orange collar palm
Arenga engleri	dwarf sugar palm, Formosa palm
Astrocaryum mexicanum	Chocho, Lancetilla, Chichon
Bactris gasipaes	peach palm, peijibaye
Burretiokentia species	
Calamus caryotoides	fishtail lawyer's cane
Calyptrocalyx hollrungii	Hollrung's palm
Carpentaria acuminata	carpentaria palm
Caryota species	
Ceroxylon alpinum	wax palm
Ceroxylon quindiuense	quindio wax palm
Ceroxylon ventricosum	
Chamaedorea species	
Chambeyronia lepidota	
Chambeyronia macrocarpa	flamethrower
Cryosophila stauracantha	
Dictyosperma album	hurricane palm, princess palm
Dypsis baronii	baron palm
Dypsis cabadae	cabada palm
Dypsis decaryi	triangle palm
Dypsis lastelliana	redneck palm
Dypsis lutescens	areca palm, butterfly palm
Euterpe edulis	cabbage palm
Euterpe oleracea	Assai palm
Gaussia species	
Guihaia argyrata	Guangxi palm
Hedyscepe canterburyana	umbrella palm, big mountain palm
Heterospathe elata	sagisi palm
Heterospathe negrosensis	
Howea species	
Hydriastele microspadix	
Hyophorbe indica	
Kentiopsis oliviformis	
Laccospadix australasicus	Atherton palm
Licuala peltata	elegant palm
Normanbya normanbyi	black palm
Oncosperma tigillarium	nibung palm
Phoenix roebelenii	pygmy date palm
Pinanga javana	
Pritchardia species	
Ptychosperma elegans	solitaire palm
Ptychosperma macarthurii	Macarthur palm
Ravenea glauca	
Ravenea hildebrandtii	
Rhapidophyllum hystrix	needle palm
Rhapis species	
Rhopaloblaste elegans	
Rhopalostylis species	
Sabal minor	dwarf blue palmetto
Satakentia liukiuensis	satake palm
Schippia concolor	silver pimento palm
Thrinax morrissii	brittle thatch palm
Thrinax parviflora	broom palm
Zombia antillarum	zombie palm

Full Shade

Botanical Name	Common Name
Calamus species	
Chamaedorea species	
Cryosophila guagara	guagara palm
Johannesteijsmannia altifrons	joey palm, diamond joey palm, umbrella leaf palm
Licuala species	
Linospadix monostachya	walking stick palm
Livistona chinensis	Chinese fan palm
Lytocaryum species	
Pinanga coronata	Coronata palm, ivory cane palm

Licuala peltata

Reference

Glossary

The list below describes most of the key terms used in this book. These terms describe many of the characteristics of palms as well as the most important related gardening terms.

Adventitious roots. Roots that form on the trunk of the tree. All palms produce adventitious roots.

Acidic. A pH value below 7.

Alkaline. A pH value above 7.

Armed. With spines.

Bipinnate. A leaf that has divisions that are once or several times compound.

Bract. A modified leaf structure on the trunk of the palm.

Canopy. The overhanging branches of a tree.

Chlorosis. Yellowing of palm fronds due to nutritional deficiencies.

Clumping palm. A palm that forms multiple stems or trunks.

Cold hardiness. The ability to survive the winter cold in a region.

Crownshaft. A smooth, sometimes shiny, area that separates the trunk from the fronds in some palms.

Cultivar. A variety or form that is known only in cultivation.

Desiccation. Drying out of fronds from drought or wind.

Frond. The leaf of the palm.

Genus. A group of related species, usually categorized by similarity or origin. The Cuban royal palm is a member of the *Roystonea* genus. Plural is genera.

Glaucous. Blue or blue-gray color on leaves covered with a waxy layer.

Inflorescence. The flower stalk.

Inflorescence bract. A boat-shaped bract that covers the inflorescence.

Leaf base. The wide area of the leaf, or frond, that attaches to the trunk.

Leaf scar. The conspicuous remnant of a fallen frond on the trunk.

Monocotylon. A flowering plant, including palms, grasses and orchids, that produces a single leaf from a seed.

Naturalize. To adapt to and spread into a non-native habitat.

Necrotic. Dead or dying plant tissue.

Palmate. Used to describe fan palm leaves.

Petiole. The primary stalk supporting the individual leaves of the palm.

pH. A measure of the acidity or alkalinity. Usually refers to the soil or water in gardening.

Pinnate. Used to describe feather-shaped palm leaves.

Plumose. Feather-shaped palm leaves that form on various planes, giving the frond a fuller look.

Spathe. A bract that subtends a flower grouping.

Species. A distinct entity, or type of palm, in a genus. The species name follows the genus. *Roystonea regia,* for example, is the correct genus and species for the Cuban royal palm.

Subtropical. Located between the Tropic latitudes and Temperate latitudes.

Spine. Sharp, rigid, pointed protrusions on various parts of some palms.

Stilt roots. Aerial roots that form at the base of the trunk to support the tree. Palms with stilt roots seem to walk on top of their surrounding soil.

Unarmed. Free of spines.

Translocate. The ability to move nutrients from one area of the palm to another.

Trunk. The main stem of the palm.

Variegated. Marked with streaks of different colors.

Pronunciation Guide

Although botanical names for palms are based on Latin and Greek, pronunciations differ throughout the world. This pronunciation guide is based on English spoken in the United States of America.

a–as in hat
ay–as in say
e–as in bed
ee–as in week
i–as in win
o–as in row
oo–as in boot
ow–as in snow
u–as in must
uh–pronounced like "the"

Botanical Name	Pronunciation
Acoelorrhaphe wrightii	a-see'-lo-RAY-fee RYT-ee-eye
Acrocomia aculeata	ak-ro-CO-mee-a a-kyoo-lee-AHT-a
Adonidia merrillii	ad-o-NID-ee-a mer-RIL-lee-eye
Aiphanes aculeata	EYE-fa-neez a-kyoo-lee-AHT-a
Aiphanes minima	EYE-fa-neez MIN-i-muh
Allagoptera arenaria	al'-la-GAHP-t'ra ar-en-NAY-ree-uh
Allagoptera campestris	al'-la-GAHP-t'ra kam-PES-tris
Archontophoenix alexandrae	ahr-kont'-o-FEE-nix a-lek-ZAN-dree
Archontophoenix cunninghamiana	ahr-kont'-o-FEE-nix kun'-ning-ham-ee-AHN-a
Archontophoenix maxima	ahr-kont'-o-FEE-nix MAX-i-ma
Archontophoenix myolensis	ahr-kont'-o-FEE-nix my-o-LEN-sis
Archontophoenix purpurea	ahr-kont'-o-FEE-nix poor-POOR-ee-a
Archontophoenix tuckeri	ahr-kont'-o-FEE-nix TUK-'-ry
Areca catechu	a-REEK-a KAT-e-choo
Areca concinna	a-REEK-a kahn-SIN-na
Areca ipot	a-REEK-a EE-paht
Areca triandra	a-REEK-a try-AN-dra
Areca vestiaria	a-REEK-a ves-tee-AHR-ee-a
Arenga engleri	a-RENG-a ENG-l'r-eye
Astrocaryum mexicanum	ass-tro-KAR-ee-um mex-i-KAN-um
Attalea cohune	at-TAL-ee-a ko-HOO-nee
Bactris gasipaes	BAK-tris GAS-i-peez
Beccariophoenix madagascariensis	bek-kahr'-ee-o-FEE-nix mad'-a-gas-kar'-ee-EN-sis
Bismarckia nobilis	bis-MARK-ee-a NO-bil-is
Borassus aethiopum	bo-RAS-sus ee-thee-O-pee-um
Borassus flabellifer	bo-RAS-sus fla-BEL-li-f'r
Brahea aculeata	bra-HAY-a a-kyoo-lee-AHT-a
Brahea armata	bra-HAY-a ahr-MAHT-a
Brahea brandegeei	bra-HAY-a bran-DEJ-ee-eye
Brahea decumbens	bra-HAY-a de-COME-bins
Brahea dulcis	bra-HAY-a DOOL-sis
Brahea edulis	bra-HAY-a ED-you-lis
Brahea nitida	bra-HAY-a ni-TI-da
Brahea moorei	bra-HAY-a MOR-ee-eye
Brahea pimo	bra-HAY-a PEE-mo
Burretiokentia hapala	bur-ret'-ee-o-KENT-ee-a ha-PAH-la
Burretiokentia viellardii	bur-ret'ee-o-KENT-ee-a vee-ai-YAHR-dee-eye
Butia archeri	BYOO-tee-a ARCH-'-ry
Butia capitata	BYOO-tee-a kap-i-TAHT-a
Butia eriospatha	BYOO-tee-a er'-ee-o-SPAITH-a
Butia microspadix	BYOO-tee-a myk-ro-SPAI-dix
Butia paraguayensis	BYOO-tee-a pah'-ra-gwah-YEN-sis
Butia yatay	BYOO-tee-a yah-TAY
Calamus australis	KAL-a-mus aw-STRAL-iss
Calamus caryotoides	KAL-a-mus kar-ee-o-TOI-deez
Calamus ciliaris	KAL-a-mus sil-ee-AIR-is
Calamus moti	KAL-a-mus MO-tee
Calamus muelleri	KAL-a-mus MYOO-l'-ry
Calamus ornatus	KAL-a-mus or-NAHT-us
Calamus radicalis	KAL-a-mus rad-i-KAL-us
Calamus reyesianus	KAL-a-mus ray-see-AN-us
Calamus viminalis	KAL-a-mus vim-in-AL-is
Calyptrocalyx hollrungii	kuh-lip-tro-KAL-ix hul-RUNG-ee-eye
Carpentaria acuminata	kahr-pen-TAHR-ee-a a-kyoo'-mi-NAHT-a
Carpoxylon macrospermum	kar-POX-i-lon mak-ro-SPERM-um
Caryota maxima	kar-ee-O-tah MAX-i-ma
Caryota mitis	kar-ee-O-tah MYT-iss
Caryota no	kar-ee-O-tah NO
Caryota obtusa	kar-ee-O-tah ob-TOO-suh
Caryota ochlandra	kar-ee-O-tah ok-LAN-dra
Caryota rumphiana	kar-ee-O-tah rump-fee-AHN-a
Caryota urens	kar-ee-O-tah YOO-renz
Ceroxylon alpinum	se-RAHX-i-lahn al-PYN-um
Ceroxylon quindiuense	se-RAHX-i-lahn keen-dee'-oo-EN-see
Ceroxylon ventricosum	se-RAHX-i-lahn ven-tri-KOS-um
Chamaedorea cataractarum	kam-mee-DOOR-ee-a kat-a-RAK-ta-rum
Chamaedorea costaricana	kam-mee-DOOR-ee-a kos-ta-REEK-a-na
Chamaedorea elegans	kam-mee-DOOR-ee-a eh-LAH-gahnz
Chamaedorea ernestii-augusti	kam-mee-DOOR-ee-a er-nest'-ee-aw-GUS-tee-eye
Chamaedorea fragrans	kam-mee-DOOR-ee-a FRA-granz
Chamaedorea glaucifolia	kam-mee-DOOR-ee-a gluh-ki-FOL-ee-uh
Chamaedorea hooperiana	kam-mee-DOOR-ee-a hoop-'r-ee-AHN-a
Chamaedorea klotzschiana	kam-mee-DOOR-ee-a klaht-shee-AHN-a
Chamaedorea metallica	kam-mee-DOOR-ee-a me-TAL-lik-a
Chamaedorea microspadix	kam-mee-DOOR-ee-a myk-ro-SPAI-dix
Chamaedorea plumosa	kam-mee-DOOR-ee-a ploo-MO-sa
Chamaedorea pochutlensis	kam-mee-DOOR-ee-a po-choot-LEN-sis
Chamaedorea seifrizii	kam-mee-DOOR-ee-a sef-FREEZ-ee-eye
Chamaedorea stolonifera	kam-mee-DOOR-ee-a stol-o-NIF-er-uh
Chamaedorea radicalis	kam-mee-DOOR-ee-a rad-i-KAL-is
Chamaedorea woodsoniana	kam-mee-DOOr-ee-a wood-son'-ee-AHN-a
Chamaerops humilis	kam-MEE-rahps HYOO-mil-lis

Botanical Name	Pronunciation
Chambeyronia lepidota	shahm-bai-RO-nee-a lep-i-DOAT-a
Chambeyronia macrocarpa	shamb-bai-RO-nee-a mak-ro-KAHRP-a
Clinostigma exorrhizum	klyn-o-STIG-ma ex-o-RY-zum
Clinostigma gronophyllum	klyn-o-STIG-ma gro-no-FYL-lum
Clinostigma harlandii	klyn-o-STIG-ma hahr-LAN-dee-eye
Clinostigma ponapense	klyn-o-STIG-ma po-na-PEN-see
Clinostigma samoense	klyn-o-STIG-ma sah-mo-EN-see
Clinostigma savoryanum	klyn-o-STIG-ma sa-vor'-ee-AHN-um
Coccothrinax argentata	kok'-ko-THRY-nax ahr-jen-TAHT-a
Coccothrinax argentea	kok'-ko-THRY-nax ahr-JEN-tee-a
Coccothrinax barbadensis	kok'-ko-THRY-nax bahr-ba-DEN-sis
Coccothrinax borhidiana	kok'-ko-THRY-nax bor-hid-ee-AN-uh
Coccothrinax crinita	kok'ko-THRY-nax kri-NEET-a
Coccothrinax cupularis	kok'ko-THRY-nax koop-yoo-LAR-is
Coccothrinax ekmanii	kok'ko-THRY-nax ek-MAHN-ee-eye
Coccothrinax fragrans	kok'ko-THRY-nax FRAI-granz
Coccothrinax hioramii	kok'ko-THRY-nax hee-o-RAH-mee-eye
Coccothrinax inaguensis	kok'ko-THRY-nax in-a-GWEN-sis
Coccothrinax jamaicensis	kok'ko-THRY-nax ja-ma-KEN-sis
Coccothrinax miraguama	kok-ko-THRY-nax meer-a-GWAHM-a
Coccothrinax munizii	kok'ko-THRY-nax mun-EE-zee-eye
Coccothrinax proctorii	kok'ko-THRY-nax proc-TOR-ee-eye
Coccothrinax spissa	kok'ko-THRY-nax SPEE-suh
Cocos nucifera	KO-kos noo-SIF-e-ra
Copernicia alba	ko-p'r-NIS-ee-a AHL-ba
Copernicia baileyana	ko-p'r-NIS-ee-a bay-lee-YAHN-a
Copernicia glabrescens	ko-p'r-NIS-ee-a gla-BRES-senz
Copernicia hospita	ko-p'r-NIS-ee-a HAHS-pi-ta
Copernicia macroglossa	ko-p'r-NIS-ee-a mak-ro-GLAHS-sa
Copernicia prunifera	ko-p'r-NIS-ee-a proo-NIF-e-ra
Corypha umbraculifera	ko-RYF-a um-brak-yoo-LIF-e-ra
Corypha utan	ko-RYF-a OO-tan
Cryosophila guagara	KRY-o-sahf-il-a gwah-GAHR-a
Cryosophila stauracantha	KRY-o-sahf-il-a stow-ra-KANTH-a
Cryosophila warscewiczii	KRY-o-sahf-il-a wahrs-se-WIK-zee-eye
Cryosophila williamsii	KRY-o-sahf-il-a WIL-yam-see-eye
Cyrtostachys renda	seer-toe-STAIK-iss REN-da
Dictyosperma album	dik-tee-o-SPURM-a AL-bum
Dypsis ambositrae	DIP-sis ahm-bo-SOT-ree
Dypsis baronii	DIP-sis ba-RO-nee-eye
Dypsis cabadae	DIP-sis ka-BAH-dee, KAB-a-dee
Dypsis decaryi	DIP-sis de-KAHR-ee-eye
Dypsis decipiens	DIP-sis de-SIP-ee-enz
Dypsis lastelliana	DIP-sis las-tel'-ee-AHN-a
Dypsis leptocheilos	DIP-sis lep-toe'-KAI-los
Dypsis lutescens	DIP-sis loo-TES-senz
Dypsis madagascariensis	DIP-sis mad-uh-gas-car-ee-EN-sis
Elaeis guineensis	e-LEE-iss gin-ee-EN-sis, gin-ay-EN-sis
Elaeis oleifera	e-LEE-iss o-lee-IF-'-ra
Euterpe edulis	yoo-TURP-ee ED-yoo-lis
Euterpe oleracea	yoo-TURP-ee o-l'-RAI-see-a

Botanical Name	Pronunciation
Gastrococos crispa	gas-tro-KO-kos KRIS-pa
Gaussia attenuata	GAWS-see-a at-ten'-yoo-AHT-a
Gaussia maya	GAWS-see-a MAH-a
Gaussia princeps	GAWS-see-a PRIN-seps
Guihaia argyrata	gwee-HY-ya ahr-jee-RAIT-a
Hedyscepe canterburyana	hed-i-SEEP-ee kant-'r-bur-ee-AHN-a
Heterospathe elata	het'-'ro-SPAITH-ee ee-LAHT-a
Heterospathe negrosensis	het'-'ro-SPAITH-ee neg-ro-SEN-sis
Howea belmoreana	HOW-ee-a bel-mor'-ee-AHN-a
Howea forsteriana	HOW-ee-a for-ster-ee-AHN-a
Hydriastele microspadix	hy-dree-ASS-te-lee myk-ro-SPA-dix
Hyophorbe indica	hy-o-FOR-bee IN-di-ka
Hyophorbe lagenicaulis	hy-o-FOR-bee lag'-e-ni-KAW-lis
Hyophorbe verschaffeltii	hy-o-FOR-bee ver-sha-FELT-tee-eye
Hyphaene coriacea	HY-fee-nee kor-ee-AI-see-a
Hyphaene dichotoma	HY-fee-nee die-ko-TO-muh
Hyphaene petersiana	HY-fee-nee peet-'rz-ee-AHN-a
Hyphaene thebaica	HY-fee-nee thee-BAI-i-ka
Johannesteijsmannia altifrons	yo-hahn-nes-tysh-MAHN-nee-a AL-tee-fronz
Jubaea chilensis	joo-BAY-a chi-LEN-sis
Jubaeopsis caffra	joo-bay-AHP-sis KAF-ra
Kentiopsis oliviformis	kent-ee-AHP-sis o-liv'-i-FOR-mis
Kentiopsis piersoniorum	kent-ee-AHP-sis peer-sahn-ee-OR-um
Kentiopsis pyriformis	kent-ee-AHP-sis py-ri-FOR-mis
Laccospadix australasicus	lak-ko-SPAI-dix aws-tra-LAI-zi-ka
Latania loddigesii	la-TAN-ee-a lo-di-GAI-zee-eye
Latania lontaroides	la-TAN-ee-a lon-ta-ROY-deez
Latania verschaffeltii	la-TAN-ee-a ver-sha-FELT-ee-eye
Lepidorrhachis mooreana	lep-i-do-RAK-is mor-ee-AN-uh
Licuala cordata	li-KWAH-la kor-DAHT-a
Licuala grandis	li-KWAH-la GRAN-dis
Licuala orbicularis	li-KWAH-la or-bik'-yoo-LAR-iss
Licuala peltata	li-KWAH-la pel-TAHT-a
Licuala ramsayi	li-KWAH-la RAM-say-eye
Licuala spinosa	li-KWAH-la spin-O-suh
Linospadix monostachya	lyn-o-SPAI-dix mo-no-STAIK-ee-a
Livistona australis	li-vi-STON-a aw-STRAL-iss
Livistona benthamii	li-vi-STON-a ben-THAM-ee-eye
Livistona chinensis	li-vi-STON-a chi-NEN-sis
Livistona decora	li-vi-STON-a de-KOOR-uh
Livistona drudei	li-vi-STON-a DROO-dee-eye
Livistona inermis	li-vi-STON-a i-NER-mis
Livistona jenkinsiana	li-vi-STON-a jen-kin-see-AN-uh
Livistona mariae	li-vi-STON-a MAHR-ee-eye
Livistona muelleri	li-vi-STON-a MYOOL-l'-ry
Livistona saribus	li-vi-STON-a sah-REE-bus
Lodoicea maldivica	lo-DOYS-ee-a mal-DIV-i-ka
Lytocaryum hoehnei	lyt-o-KAHR-ee-um HURN-ee-eye
Lytocaryum weddellianum	lyt-o-KAHR-ee-um wed-del'-ee-AHN-um
Medemia argun	me-DEM-ee-a ahr-GOON

Botanical Name	Pronunciation
Metroxylon amicarum	me-TRAHX-i-lahn am-i-KAHR-um
Metroxylon sagu	me-TRAHX-i-lahn SAH-goo
Metroxylon vitiense	me-TRAHX-i-lahn vit-ee-EN-se
Metroxylon warburgii	me-TRAHX-i-lahn wahr-BURG-ee-eye
Nannorrhops ritchiana	NAN-o-rahps rich-ee-AHN-a
Neoveitchia storckii	nee-o-VEECH-ee-a STORK-ee-eye
Normanbya normanbyi	nor-MAN-bee-a nor-MAN-bee-eye
Nypa fruticans	NY-pa FROOT-i-kanz
Oncosperma tigillarium	ahnk-o-SPURM-a tij-i-LAHR-ee-um
Parajubaea cocoides	par'-a-joo-BEE-a kok-KOI-deez
Parajubaea torallyi	par'-a-joo-BEE-a toe-RAL-lee-eye
Phoenix canariensis	FEE-nix ka-nar'-ee-EN-sis
Phoenix dactylifera	FEE-nix dac-ti-LIF-e-ra
Phoenix loureiroi	FEE-nix loo-RAI-ry
Phoenix pusilla	FEE-nix poo-SIL-la
Phoenix reclinata	FEE-nix rek-li-NAHT-a
Phoenix roebelenii	FEE-nix ro-be-LEN-ee-eye
Phoenix rupicola	FEE-nix roop-i-KO-la
Phoenix sylvestris	FEE-nix sil-VES-tris
Phoenix theophrasti	FEE-nix thee-o-FRAS-tee
Pinanga caesia	pin-NAING-a KAY-see-a
Pinanga coronata	pi-NAING-a kor-o-NAHT-a
Pinanga javana	pi-NAING-a ja-VAHN-a
Pinanga philippinensis	pi-NAING-a fil'-i-pi-NEN-sis
Pritchardia affinis	prit-CHAHRD-ee-a AF-fi-nis
Pritchardia arecina	prit-CHAHRD-ee-a ar-e-SEE-na
Pritchardia beccariana	prit-CHAHRD-ee-a bek-kahr'-ee-AHN-a
Pritchardia hillebrandii	prit-CHAHRD-ee-a hil-l'-BRAN-dee-eye
Pritchardia kaalae	prit-CHAHRD-ee-a kah-AH-lee
Pritchardia lanaiensis	prit-CHAHRD-ee-a la-nah'-ee-EN-sis
Pritchardia lowreyana	prit-CHAHRD-ee-a low-ray-AN-a
Pritchardia martii	prit-CHAHRD-ee-a MAHRT-ee-eye
Pritchardia pacifica	prit-CHAHRD-ee-a pa-SIF-i-ka
Pritchardia remota	prit-CHAHRD-ee-a ree-MOT-a
Pritchardia thurstonii	prit-CHAHRD-ee-a thur-STONE-ee-eye
Pseudophoenix sargentii	soo-doe-FEE-nix sahr-JEN-tee-eye
Pseudophoenix vinifera	soo-doe-FEE-nix vi-NIF-e-ra
Ptychosperma elegans	ty-ko-SPURM-a EL-e-ganz
Ptychosperma macarthurii	ty-ko-SPURM-a mak-AHRTH-'ree-eye
Raphia farinifera	RAF-ee-a far-i-NIF-'-ra
Raphia hookeri	RAF-ee-a HOOK-'-ry
Ravenea glauca	rav-e-NAI-a GLAW-ka
Ravenea hildebrandtii	rav-e-NAI-a hil-d'-BRANT-tee-eye
Ravenea rivularis	rav-e-NAI-a riv-yoo-LAR-iss
Rhapidophyllum hystrix	rap-I-doe-FYL-lum HISS-trix
Rhapis excelsa	RAP-is ek-SEL-sa
Rhapis humilis	RAP-is HYOO-mi-lis
Rhapis subtilis	RAP-is SUB-ti-liss
Rhopaloblaste elegans	rop'-a'lo-BLAS-tee EL-e-ganz
Rhopalostylis baueri	ROPE-a-lo-STY-liss BOW-er-eye
Rhopalostylis cheesemanii	ROPE-a-lo-STY-liss cheez-MAHN-ee-eye

Botanical Name	Pronunciation
Rhopalostylis sapida	rop-a'lo-STY-liss SAP-i-da
Roystonea altissima	roy-STON-ee-a al-TIS-si-ma
Roystonea borinquena	roy-STON-ee-a bo-reenk'-AIN-a
Roystonea oleracea	roy-STON-ee-a o-le-RAI-see-a
Roystonea princeps	roy-STON-ee-a PRIN-seps
Roystonea regia	roy-STON-ee-a REE-jee-a
Sabal bermudana	SAI-bahl ber-myoo-DAHN-a
Sabal causiarum	SAI-bahl kow-see-AHR-um
Sabal domingensis	SAI-bahl doe-meen-GEN-sis
Sabal etonia	SAI-bahl ee-TOE-nee-a
Sabal louisiana	SAI-bahl loo-eez-ee-AN-a
Sabal mexicana	SAI-bahl mex-i-KAHN-a
Sabal minor	SAI-bahl MY-nor
Sabal palmetto	SAI-bahl pahl-MET-toe
Sabal rosei	SAI-bahl RO-zee-eye
Sabal uresana	SAI-bahl oo-re-SAHN-a
Satakentia liukiuensis	saht-a-KENT-ee-a lee-ook'-ee-oo-EN-sis
Schippia concolor	SHIP-pee-a KAHN-kul-or
Serenoa repens	ser-e-NO-a REP-enz
Syagrus amara	si-AG-rus a-MAHR-a
Syagrus cearensis	si-AG-rus see-uh-REN-sis
Syagrus comosa	si-AG-rus ko-MO-sa
Syagrus coronata	si-AG-rus ko-ro-NAHT-a
Syagrus pseudococos	si-AG-rus soo-doe-KO-kos
Syagrus romanzoffiana	si-AG-rus ro-man-zof-ee-AHN-a
Syagrus sancona	si-AG-rus san-KO-na
Syagrus schizophylla	si-AG-rus skits-o-FYL-la
Thrinax excelsa	THRY-nax ek-SEL-sa
Thrinax morrisii	THRY-nax mor-RIS-ee-eye
Thrinax radiata	THRY-nax ray-dee-AH-ta
Thrinax parviflora	THRY-nax pahr-vi-FLO-ra
Trachycarpus fortunei	traik-ee-KAHRP-us for-TOON-ee-eye
Trachycarpus latisectus	traik-ee-KAHRP-us lat-i-SEK-tus
Trachycarpus martianus	traik-ee-KAHRP-us mahrt-ee-AHN-us
Trachycarpus nanus	traik-ee-KAHRP-us NAN-us
Trachycarpus takil	traik-ee-KAHRP-us TAH-keel
Trithrinax brasiliensis	try-THRY-nax bra-zil'-ee-EN-sis
Trithrinax campestris	try-THRY-nax kam-PES-tris
Veitchia arecina	VEECH-ee-uh ar-e-SEE-na
Veitchia filifera	VEECH-ee-uh fi-LIF-e-ra
Veitchia joannis	VEECH-ee-uh jo-AN-nis
Veitchia winin	VEECH-ee-uh WINE-in
Verschaffeltia splendida	ver-sha-FELT-ee-uh splen-DEED-a
Washingtonia filifera	wahsh-ing-TO-nee-uh fi-LIF-e-ra
Washingtonia robusta	wahsh-ing-TO-nee-uh ro-BUS-ta
Welfia regia	WELF-ee-uh REE-jee-a
Wodyetia bifurcata	wod-YET-ee-uh by-foor-KAHT-a
Zombia antillarum	ZAHM-bee-uh an-til-LAHR-um

Public gardens filled with palms

Below is a list of gardens that have excellent palm collections. At many of these locations, you can view the palms that were photographed for this book. These gardens will allow you to compare a variety of characteristics in a large collection of palms, providing you with inspiration and ideas for your own landscape.

United States

Fairchild Tropical Botanic Garden
10901 Old Cutler Road
Coral Gables, FL 33156
(305) 667-1651
http://www.fairchildgarden.org

Foster Botanical Garden
50 North Vineyard Boulevard
Honolulu, Hawaii 96817
(808) 522-7066
http://www.co.honolulu.hi.us/parks/hbg/fbg.htm

Fullerton Arboretum
1900 Associated Road
Fullerton, CA 92831
(714) 278-3579
http://www.arboretum.fullerton.edu

Gizella Kopsick Palm Arboretum
North Shore Drive
St. Petersburg, FL
727) 893-7335
http://www.stpete.org/palm.htm

Hawaii Tropical Botanical Garden
27-717 Old Mamalahoa Highway
Papaikou, HI 96781
808-964-5233
http://www.htbg.com

Hoomaluhia Botanical Garden
45-680 Luluku Road
Kaneohe, HI 96744
(808) 233-7323
http://www.co.honolulu.hi.us/parks/hbg/hmbg.htm

Huntington Library, Art Collections, & Botanical Gardens
1151 Oxford Road
San Marino, CA 91108
(626) 405-2100
http://www.huntington.org

Koko Crater Botanical Garden
Inside Koko Crater
Honolulu, HI 96825
(808) 522-7060
http://www.co.honolulu.hi.us/parks/hbg/kcbg.htm

Los Angeles County Arboretum & Botanic Garden
301 N. Baldwin Avenue
Arcadia, CA 91007
(626) 821-3222
http://www.arboretum.org

Lotusland
695 Ashley Road
Santa Barbara, CA 93108
(805) 969-9990
http://www.lotusland.org

Lyon Arboretum
3860 Manoa Road
Honolulu, Hawaii 96822
(808) 988-0456
http://wwwdev.hawaii.edu/lyonarboretum

Marie Selby Botanical Garden
811 South Palm Avenue
Sarasota, FL 34236
(941) 366-5731
http://www.selby.org

Mildred E. Mathias Botanical Garden
University of California, Los Angeles
Los Angeles, CA 90095
310-825-1260
http://www.botgard.ucla.edu

Missouri Botanical Garden
4344 Shaw Boulevard
St. Louis, MO 63110
(314) 577-9400
http://www.mobot.org

Myakka River State Park
13207 S.R. 72
Sarasota, FL 34241
(941) 361-6511
http://www.myakkariver.org

Nani Mau Gardens
421 Makalika Street
Hilo, HI 96720
(808) 959-3500
http://www.nanimau.com

National Tropical Botanical Garden
3530 Papalina Road
Kalaheo, HI 96741
(808) 332-7324
Locations throughout Hawaii
http://ntbg.org

The New York Botanical Garden
Bronx River Parkway
Bronx, NY 10458
(718) 817-8700
http://www.nybg.org

Palm Canyon
Palm Springs, CA
(800) 790-3398
http://www.palmsprings.com/points/canyon/index.html

Panaewa Rainforest Zoo and Garden
Hilo, HI
808-959-9233
http://www.hilozoo.com

Quail Botanical Gardens
230 Quail Gardens Drive
Encinitas, CA 92024
(760) 436-3036
http://www.qbgardens.org

San Diego Zoo
2920 Zoo Drive
San Diego, CA 92100
(619) 234-3153
http://www.sandiegozoo.org

Virginia Robinson Gardens
Beverly Hills, CA 90210
(310) 276-5367
http://parks.co.la.ca.us/virginia_gardens.html

Wahiawa Botanical Garden
1396 California Avenue
Wahiawa, HI 96786
Phone: (808) 621-7321
http://www.co.honolulu.hi.us/parks/hbg/wbg.htm

Waimea Valley Audubon Center
59-864 Kamehameha Highway
Haleiwa, Hawaii, 96712
(808) 638-9199
http://www.audubon.org/local/sanctuary/Brochures/Waimea.html

World Botanical Gardens
19 Mile Marker, Highway 19
Umauma, HI
http://www.wbgi.com

Australia

Brisbane Botanic Gardens
Mt Coot-tha Road
Brisbane, Queensland
07 3403 8888
http://www.brisbane.qld.gov.au/BCC:BOTANICG::pc=PC_1346

Townsville Palmetum
Nathan Street
Townsville Qld 4810
07 4727 8330
http://www.townsville.qld.gov.au/services/parks/palmetum.asp

England

Royal Botanic Gardens, Kew
Richmond, Surrey TW9 2AB
020 8332 5655
http://www.rbgkew.org.uk

Thailand

Nong Nooch Tropical Garden
Pattaya, Thailand
(038) 429-321

Maldives

Pamplemousses Gardens
Near Point Louis

Singapore

Singapore Botanic Gardens
1 Cluny Road
Singapore 259569
(65) 6471 7361
http://www.sbg.org.sg

Costa Rica

La Selva Biological Station
Apartado 676-2050
San Pedro, Costa Rica.
Phone: (506) 766 6 565
http://www.ots.ac.cr/en/laselva

Las Cruces-Wilson Botanical Garden
Apartado 73-8257
San Vito, Coto Brus, Costa Rica.
Phone: (506) 773-4004
http://www.ots.ac.cr/en/lascruces

Sri Lanka

Peradeniya Royal Botanic Gardens

India

Lalbagh Botanical Garden
Bangalore
http://horticulture.kar.nic.in/lalbagh.htm

Panama

Summit Botanical Garden
Soberania National Park
Panama City

Venezuela

Caracas Botanical Garden
Corp de Turismo de Venezuela
Av Bolivar, Parque Central - Torre
Oeste Piso 37, El Conde
Caracas 1010
+58 (0) 2 5741968 or +58 (0) 2 5742124

Brazil

Rio de Janeiro Botanical Gardens
Rua Jardim Botanico 1008,
Rio de Janeiro
Tel: 021-294-6012
http://www.jbrj.gov.br

Puerto Rico

University of Puerto Rico Botanical Garden
Road 1 Kilometer 12
San Juan, PR 00931
USA
Phone: 787-763-4408
http://www.upr.clu.edu/jardin/JardiBot.htm

Belize

Belize Botanic Gardens
P.O. Box 180
San Ignacio, Cayo, Belize
(501) 824-3101
http://www.belizebotanic.org

Dominica

Dominica Botanic Gardens
Roseau, Dominica
http://www.da-academy.org/dagardens.html

Bibliography

Barrow, Sasha. 1998. "A Revision of Phoenix." Reprinted from *Kew Bulletin* 53(3). London.

Broschat, Timothy K., and Alan W. Meerow. 2000. *Ornamental Palm Horticulture.* University Press of Florida, Gainesville, Florida.

Dransfield, John, and Henk Beentje. 1995. *The Palms of Madagascar.* Board of Trustees of the Royal Botanic Gardens, Kew, London.

Ellison, Don, and Anthony Ellison. 2001. *Betrock's Cultivated Palms of the World.* Betrock Information Systems, Inc., Hollywood, Florida.

Gibbons, Martin. 1993. *Palms.* Chartwell Books, Seacaucus, New Jersey.

Henderson, Andrew, Gloria Galeano, and Rodrigo Bernal. 1995. *Palms of the Americas.* Princeton University Press, Princeton, New Jersey.

Hodel, Donald R., ed. 1998. *The Palms and Cycads of Thailand.* Allen Press, Lawrence, Kansas.

Hodel, Donald R. 1992. *Chamaedorea Palms: The Species and their Cultivation.* Allen Press, Lawrence, Kansas.

Jones, Davis L. 1995. Pa*lms Throughout the World.* Smithsonian Institution Press, Washington D.C.

Leaser, David E. 2005. *Palm Trees: A Story in Photographs.* Westwood Pacific Publishing, Los Angeles, California.

Leaser, David E. 2007. *Growing Palm Trees in Hawaii and other Tropical Locations.* Mutual Publishing LLC, Honolulu, Hawaii.

Meerow, Alan W. 2005. *Betrock's Cold Hardy Palms.* Betrock Information Systems, Inc., Hollywood, Florida.

Meerow, Alan W. 2006. *Betrock's Landscape Palms.* Betrock Information Systems, Inc., Hollywood, Florida.

Osborn, B., T. Reynoso & G. Stein. 2000. *Palms for Southern California – A Quick Reference Guide to Palms.* Third Edition. The Palm Society of Southern California.

PRINCIPES (later *PALMS*). Journal of the International Palm Society. 1956-2006. Vols. 1-50. Lawrence, Kansas.

Riffle, Robert Lee, and Paul Craft. 2003. *An Encyclopedia of Cultivated Palms.* Timber Press, Portland, Oregon.

Romney, David H. 1997. *Growing Coconuts in South Florida.* Homestead, Florida.

Sunset. 2007. *Western Garden Book.* Sunset Publishing Corp., Menlo Park, California.

Uhl, Natalie W., and John Dransfield. 1987. *Genera Palmarum.* L. H. Bailey Hortorium and International Palm Society. Lawrence, Kansas.

Warren, William. 1997. *The Tropical Garden.* Thames and Hudson, Inc., New York, New York.

Many thanks!

Many thanks to the following for their knowledge, skills and support of my work on this book.

Members of the International Palm Society provided accurate, up-to-date information on the culture and care of palms. Thanks to Scott Zona, Don Hodel, Randy Moore, Jim Wright, Billy Olson, Karen and Dean Piercy, and Bo Lundkvist. Tobias Spanner of *rarepalmseeds.com* provided excellent photos of palms in their native habitat.

I visited several of the finest palm gardens to photograph their collections. Thank you to Joshlyn Sand and Jonel Smith from the Honolulu Botanical Gardens, Mike Maunder and the staff of the Fairchild Tropical Botanic Garden, and Jim Folsom, Dani Rudeen and Melanie Thorpe from the Huntington Library, Art Collections, & Botanical Gardens.

The Montgomery Botanical Center, a nonprofit institution devoted to advancing the science of tropical botany, displays an impressive collection of rare and exotic palms. Thanks to Patrick Griffith, Larry Noblick, Laurie Danielson and the staff at MBC, I was invited to photograph their collection for this book.

Thanks also to the nurserymen, growers and landscapers who provided real-world care information, including Chris Oppenheimer from Botanics Wholesale, leading landscape architect Robert Davie, and landscaper Domingo Gonzalez.

Finally, thanks to my family and friends for their help, including Terrance Leaser, Nate Leaser, Tammy and Richard Guerra, Emi Harnden, Bob Redpath, Matt Ingebrigtsen, Heidi Wulkow and Joseph and Elizabeth Leaser.

Palm index

About the author

David Leaser has authored multipe books and journal articles about palms. His book, *Palm Trees: A Story in Photographs*, the first coffee-table book ever written on palms, has received critical acclaim from horticultural societies and leading authorities on palm trees around the world, including the Royal Botanical Garden at Kew.

Mr. Leaser authored *Growing Palm Trees in Hawaii and other Tropical Climates*, a beginner's reference to growing palms in tropical locations.

A long-time member of the International Palm Society and the International Plant Propagators Society, Mr. Leaser is in demand as a featured speaker at botanical gardens and garden centers throughout the United States, including the Los Angeles Arboretum, the San Francisco Botanical Garden, the Huntington Library, Art Collections, and Botanical Gardens and the San Diego Zoo. He lectures on a variety of topics, including "Good Palms for Small Gardens," "Landscaping with Palms," and "Seven Surprising Uses for Palms."

Mr. Leaser has also written articles for publications, including *Garden Compass Magazin*e and the *Palm Journal*, and his photos have appeared in *Sunset, Islands Magazine* and various palm-related journals and newspapers. He is frequently interviewed on palm-oriented topics by radio programs and publications, including Garden Compass Radio, Garden Talk Radio, the *Los Angeles Times,* the *Los Angeles Daily Breeze*, the *Malibu Surfside News* and the *Sunset Western Garden Book.*

A graduate of Pepperdine University, Mr. Leaser received the Pacemaker® award for excellence in journalism. He received his master's degree from the University of Southern California and also attended the University of Hawaii at Manoa. David Leaser resides in Los Angeles, California.